LHCb 上 W 玻色子的产生截面及其电荷不对称性

张文超 著

中国石化出版社

图书在版编目(CIP)数据

LHCb 上 W 玻色子的产生截面及其电荷不对称性/张文超著. —北京：中国石化出版社，2021.3
ISBN 978-7-5114-6104-9

Ⅰ. ①L… Ⅱ. ①张… Ⅲ. ①夸克－探测器－研究 Ⅳ. ①O572.33

中国版本图书馆 CIP 数据核字(2021)第 033290 号

未经本社书面授权,本书任何部分不得被复制、抄袭,或者以任何形式或任何方式传播。版权所有,侵权必究。

中国石化出版社出版发行
地址:北京市东城区安定门外大街 58 号
邮编:100011 电话:(010)57512500
发行部电话:(010)57512575
http://www.sinopec-press.com
E-mail:press@sinopec.com
北京柏力行彩印有限公司印刷
全国各地新华书店经销

*

710×1000 毫米 16 开本 10.5 印张 209 千字
2021 年 3 月第 1 版 2021 年 3 月第 1 次印刷
定价:58.00 元

前言

大型强子对撞机(LHC)是目前世界上最大和能量最高的粒子加速器,它允许物理学家在较高的能区检验标准模型,能够回答基本粒子的质量是否是由希格斯机制产生。另外,在 LHC 上可以寻找一些新物理如超对称和额外维预言的新粒子。2008 年 9 月 10 日,LHC 开始第一轮加速质子,但 9 天后两块超导磁铁连接处发生了严重的故障。由于此故障,质子—质子对撞被耽搁了 14 个月。2010 年 3 月 30 日,两束能量均为 3.5 TeV 的质子束流对撞。2012 年 4 月,束流的能量升高到 4 TeV。2015 年,束流的能量达到 6.5 TeV。LHC 上有四个对撞点,每个对撞点上均安装一个探测器,它们是 ATLAS、CMS、ALICE 和 LHCb。

本书的主要内容是介绍如何用 LHCb 探测器收集的数据测量质子—质子在质心系能量为 7 TeV 碰撞下 W 玻色子的产生截面。本书中所有的图和表格除特别说明外,均由作者本人绘制。

LHCb 是一个单臂前向探测器,其主要目的是研究 B 物理。它的赝快度(η)覆盖区间为 $1.9 < \eta < 4.9$。2010 年,LHCb 探测器收集了 7 TeV 下约 $37 pb^{-1}$ 的实验数据。在本书中,作者用 2010 年的数据测量了 $W \to \mu\nu_\mu$ 过程的微分截面和总截面。对此截面的测量可以用来检验标准模型,同时还可以减少部分子分布函数的不确定度。此外,它还可以作为 ATLAS 和 CMS 测量结果的补充。$W \to \mu\nu_\mu$ 事例的选择条件由其蒙特卡洛样本确定。在该选择条件下,实验数据中约有 26891 个 $W \to \mu\nu_\mu$ 事例,其中 $W^+ \to \mu^+ \nu_\mu$ 事例有 15030 个,$W^- \to \mu^- \nu_\mu$ 事例有 11861 个。W 玻色子产生的缪子纯度由拟合来确定,该纯度约为 79%。$W \to \mu\nu_\mu$ 过程的最终截面是在下面基准相空间中测量的:缪子的横动量大于 20GeV/c,且其赝快度区间为 $2.0 < \eta < 4.5$。在基准相空间中,$W^+ \to \mu^+ \nu_\mu$ 和 $W^- \to \mu^- \nu_\mu$ 过程的总截面分别为:

$$\sigma_{W^+ \to \mu^+ \nu_\mu} = 890.1 \pm 9.2 \pm 26.4 \pm 30.9 \text{ pb}$$

$$\sigma_{W^-\to\mu^-\bar{\nu}_\mu} = 687.0\pm 8.1\pm 19.6\pm 23.7 \text{ pb}$$

式中,不确定度第一项是统计误差,第二项是系统误差,第三项是由于数据亮度引起的误差。$W^+\to\mu^+\nu_\mu$ 和 $W^-\to\mu^-\bar{\nu}_\mu$ 过程的总截面的比值为:

$$\frac{\sigma_{W^+\to\mu^+\nu_\mu}}{\sigma_{W^-\to\mu^-\bar{\nu}_\mu}} = 1.292\pm 0.020\pm 0.002$$

式中,不确定度第一项是统计误差,第二项是系统误差。截面比值消除了由数据亮度引起的误差。

W玻色子截面和其比值的测量结果和用MSTW08及JR09部分子分布函数的次次领头阶(NNLO)理论预测结果一致。用ABKM09部分子分布函数计算得到的截面比值预言结果比实验值偏高。$W^+(W^-)$截面的测量精度约为4.7%(4.6%)。该精度比用MSTW08、ABKM09和JR09部分子分布函数计算得到的截面理论预测值精度1.8%、1.7%和2.9%(2.6%、1.6%和3.2%)大。然而,由于亮度引起的截面误差相消,且$W^+(W^-)$截面的系统误差相互关联,截面比值的不确定度主要来源于统计误差。$W^+(W^-)$总截面的比值精度为1.6%。该精度比用MSTW08及JR09部分子分布函数计算得到的比值精度3.3%和3.1%小。故对W玻色子截面比值的测量可以减少MSTW08及JR09理论预测值的误差。

本书第一章为概述。第二章对标准模型和质子-质子对撞下W玻色子截面的理论计算做了较详细的介绍。第三章和第四章对LHCb探测器以及事例重建做了较完整的描述。第五章具体陈述了W玻色子产生截面的测量。第六章是对全书的总结。

本书的原始手稿由作者一人完成,在此特别感谢两位硕士研究生车国荣和顾锦彪的帮助。由于作者水平有限,书中难免存在疏漏,如读者发现不当之处,请通过邮箱 wenchao.zhang@snnu.edu.cn 指正。

Contents

Chapter 1 Introduction ……………………………………… (1)
Chapter 2 Theoretical review ……………………………… (3)
  2.1 Standard model ……………………………………… (3)
  2.2 W Boson production at LHCb …………………… (23)
  2.3 Parton shower and hadronization ………………… (37)
  2.4 Monte carlo event generators …………………… (39)
  2.5 $\sigma_{W \to l\nu}$ and $A^l_\pm$ measurements at GPDs ………… (40)
Chapter 3 Experimental environment …………………… (44)
  3.1 LHC ……………………………………………………… (44)
  3.2 LHCb …………………………………………………… (47)
Chapter 4 Event processing at LHCb …………………… (63)
  4.1 Track reconstruction ……………………………… (63)
  4.2 Primary vertex reconstruction …………………… (69)
  4.3 Particle identification ……………………………… (71)
  4.4 LHCb trigger ………………………………………… (78)
  4.5 LHCb stripping ……………………………………… (82)
  4.6 LHCb software ……………………………………… (82)
Chapter 5 $\sigma_{W \to \mu\nu_\mu}$ measurement at LHCb …………… (85)
  5.1 Signal and background processes ………………… (85)
  5.2 Track pre-selection requirements ………………… (87)
  5.3 Pseudo-W data sample …………………………… (92)
  5.4 Candidate selection cuts …………………………… (93)
  5.5 Templates in the fit ………………………………… (99)
  5.6 Fit results …………………………………………… (114)
  5.7 Muon track detector efficiency …………………… (117)
  5.8 Muon track selection efficiency …………………… (128)
  5.9 Muon track acceptance efficiency ………………… (129)

5.10	Final state radiation	(130)
5.11	$\sigma_{W \to \mu\nu_\mu}$ determination	(131)
5.12	$\sigma_{W \to \mu\nu_\mu}$ ratio	(139)
5.13	$\sigma_{W \to \mu\nu_\mu}$ charge asymmetry	(141)
5.14	$\sigma_{W \to \mu\nu_\mu}$ theoretical predictions	(142)
5.15	Comparing LHCb results with ATLAS	(149)

Chapter 6   Conclusion .... (154)

Bibliography .... (156)

Chapter 1

Introduction

Particle physics[1] studies the nature of fundamental particles which consist of matter and anti-matter. It also studies forces between these particles. A theoretical framework called Quantum Field Theory (QFT) is applied[2]. It is a combination of Quantum Mechanics (QM) and special relativity. QFT treats a particle as an excited state of an underlying physical field, and it treats interactions between particles through the exchange of mediating particles.

QFT was first successfully applied to classical electromagnetism. This is known as Quantum Electrodynamics (QED)[3]. It describes interactions between electrically charged particles by the exchange of photons. As described in section 2.1.4, the massless photon is the consequence of the gauge symmetry of QED. QED has been called the "jewel of physics" by Richard Feynman as it predicts quantities such as the anomalous magnetic moment of the electron[4], and the Lamb shift of the energy levels of hydrogen[5].

QFT has also been applied to weak interactions. This is known as Electroweak Theory (see section 2.1.9)[6]. It describes weak interactions between particles by the exchange of W and Z bosons. As described in section 2.1.8, massive gauge bosons W and Z are the consequences of spontaneous symmetry breaking of the gauge symmetry in electroweak theory.

Furthermore, QFT has been applied to strong interactions. This is known as Quantum Chromodynamics (QCD)[7]. It describes strong interactions between particles by the exchange of gluons. As described in section 2.1.5, massless gluons are the consequences of the gauge symmetry of QCD.

An overarching framework brings these theories together and it is called the Standard Model (SM)[8]. Although the predictions of the SM are largely confirmed by experiments with good precision[9,10], there are still some inadequacies of the SM. These inadequacies include:

- Gravitation is not accounted for in the SM because it is not renormalizable as the coupling constant of gravity G has negative mass dimensions. Therefore the SM is incomplete.
- The SM can not fully account for the matter anti-matter asymmetry observed in nature. Although the SM predicts matter anti-matter asymmetry via CP violation (see section 2.1.11), it can only partially account for the total observed in nature[11].

In order to test the SM, several particle colliders have been built around the world, such as the Large Electron-Positron Collider (LEP)[12], Tevatron[13], and the Large Hadron Collider (LHC)[14]. These colliders accelerated electrons or protons to a very high energy and then let them collide. Detectors were built around collision points in order to record positions, momenta and energies of final state particles after collision. Due to the mass-energy equivalence $E=mc^2$, some massive particles such as the W, Z bosons and the top quark can be produced in high energy collisions. Today, the LHC (see section 3.1) is the highest energy collider around the world.

In this book, the cross-section measurement for the process where a W boson decays into a muon and a neutrino is described. This measurement was performed at the LHCb detector. A $W \to \mu \nu_\mu$ simulation sample is utilized to examine the characteristics of $W \to \mu \nu_\mu$ events in this measurement. Then an event selection scheme is determined based on these characteristics. This event selection scheme is applied on a data sample in order to keep signal events and suppress background events. The background events are studied with simulation samples as well as data driven samples. A fit is performed to determine how many signal events are in the data sample. A $W \to \mu\nu$ data sample is utilized to determine efficiencies of reconstructing and selecting signal events. Finally the $W^\pm$ cross-sections and their cross-section ratio and charge asymmetry are calculated.

The organization of this book is as follows. Chapter 2 reviews the theoretical background required to measure the $W \to \mu\nu_\mu$ cross-section. Chapter 3 describes the LHC acceleration system and the LHCb detector. Chapter 4 illustrates the event reconstruction procedure. Chapter 5 shows the measurement of the $W \to \mu\nu_\mu$ cross-section. Finally chapter 6 gives a brief conclusion of the $W \to \mu\nu_\mu$ cross-section measurement.

Chapter 2

Theoretical review

This chapter reviews the theoretical background for the analysis presented in this book. In section 2.1, the content of the standard model is presented. In section 2.2, the theory for W boson production in proton-proton (pp) collisions is shown. In section 2.3, the parton shower and hadronization in pp collisions are provided. In section 2.4, the Monte Carlo generators used in the analysis are described. In section 2.5, the $W^{\pm}$ cross-sections and their charge asymmetry measurements at general particle detectors are reviewed.

2.1 Standard model

The standard model includes descriptions of elementary particles as well as composite particles. These particles interact through the electromagnetic, weak and strong forces. These three forces are combined together in a $U(1) \times SU(2) \times SU(3)$ gauge theory. The electromagnetic force is described by QED. The strong force is described by QCD. A major difference between QED and QCD is the way in which the size of the force scales with energy. This leads to asymptotic freedom as well as confinements of elementary particles in mesons and baryons. Electroweak theory combines the weak and electromagnetic forces together. In this theory, W and Z bosons acquire masses through the Higgs mechanism. In order to understand this mechanism, knowledge about spontaneous symmetry breaking is needed. The particle corresponding to the scalar field utilized in the Higgs mechanism is called the Higgs boson. Quarks change their flavors in different generations through the exchange of W bosons. This is explained by a mechanism called quark mixing. CP violation is a result of the quark mixing in three generations.

2.1.1 Elementary particles

In particle physics, an elementary particle is considered to have no measurable internal structure. In the standard model of particle physics, elementary particles are classified into two groups according to their spins: the half-integer spin fermions which include quarks, leptons and their anti-particles, and the integer-spin bosons which include the $W^{\pm}$, Z, photon, gluon, and the Higgs boson.

Leptons

There are six types of leptons in total. Their respective anti-particles are anti-leptons. Leptons are grouped into three generations according to their increasing masses. In each generation there are a pair of leptons. Leptons in the first generation are the electron and electron neutrino (e^-, ν_e). In the second generation they are the muon and muon neutrino (μ^-, ν_μ). In the third generation they are the tau and tau neutrino (τ^-, ν_τ). e^-, μ^- and τ^- carry -1 electric charge while their anti-particles carry +1 charge. ν_e, ν_μ, ν_τ and their anti-particles are neutral. Charged leptons interact via electromagnetic and weak interactions while neutral leptons interact only via weak interactions. Table 2.1 shows the spins, charges and masses of leptons.

Table 2.1 The properties of leptons. The last two digits in the brackets give the uncertainties. Taken from Ref. [15]

Lepton generation	Name	Symbol	Spin	Charge(e)	Mass (MeV/c^2)
1	Electron	e	$\frac{1}{2}$	-1	0.510998928(11)
1	Electron neutrino	ν_e	$\frac{1}{2}$	0	$< 2 \times 10^{-6}$
2	Muon	μ	$\frac{1}{2}$	-1	105.6583715(35)
2	Muon neutrino	ν_μ	$\frac{1}{2}$	0	< 0.19
3	Tau	τ	$\frac{1}{2}$	-1	1776.82(16)
3	Tau neutrino	ν_τ	$\frac{1}{2}$	0	< 18.2

Quarks

There are six types of quarks in total. Each type corresponds to one flavor. The anti-particles of quarks are called anti-quarks. Anti-quarks carry opposite electric charges. Quarks are grouped into three generations. In each generation there are a pair of quarks: one is the up-type quark, which carries +2/3 electric charge, the other one is the down-type quark, which carries -1/3 electric charge.

There are three possible color charges for quarks: red, blue and green. Anti-quarks carry opposite color charges. This leads to 18 quarks and 18 anti-quarks. Due to a phenomenon known as confinement, quarks (anti-quarks) can not be isolated or directly observed. They can only be found within composite particles, such as baryons and mesons. As quarks (anti-quarks) carry electric charges as well as color charges, they can interact via electromagnetic, weak and strong interactions. Table 2.2 shows the quark properties.

Table 2.2 The properties of quarks. Taken from Ref. [15]

Quark generation	Name	Symbol	Spin	Charge(e)	Mass(MeV/c^2)
1	Up	u	$\frac{1}{2}$	$+\frac{2}{3}$	$2.3^{+0.7}_{-0.5}$
1	Down	d	$\frac{1}{2}$	$-\frac{1}{3}$	$4.8^{+0.7}_{-0.3}$
2	Charm	c	$\frac{1}{2}$	$+\frac{2}{3}$	1275 ± 25
2	Strange	s	$\frac{1}{2}$	$-\frac{1}{3}$	95 ± 5
3	Top	t	$\frac{1}{2}$	$+\frac{2}{3}$	173500 ± 1000
3	Bottom	b	$\frac{1}{2}$	$-\frac{1}{3}$	4190 ± 30

Mediators

In the standard model, the electromagnetic, weak and strong forces are mediated by elementary spin-1 bosons. In the electromagnetic interaction the mediator is a photon, γ. As described in section 2.1.4, the photon is massless. In the weak interaction the mediators are the massive $W^\pm$ and Z bosons. In the strong interaction the mediator is a massless gluon. There are 8 types of gluons carrying color charges. The masses of all particles in the standard model are believed to be created by their interactions with the Higgs boson. At the time of writing, a new particle with a mass around 125 GeV/c^2 was discovered in July, 2012 by ATLAS[16] and CMS[17] at the LHC. As this new particle behaves in many of the expected ways predicted by the standard model, it is preliminarily confirmed to be the standard model Higgs boson[18]. The mediators properties are listed in Table 2.3.

Table 2.3 The properties of mediators. Taken from Ref. [15]

Name	Symbol	Spin	Charge(e)	Mass (GeV/c^2)	Interaction mediated
Photon	γ	1	0	$< 1 \times 10^{-27}$	Electromagnetism
W boson	$W^\pm$	1	± 1	80.385 ± 0.015	Weak interaction

continue table

Name	Symbol	Spin	Charge(e)	Mass (GeV/c²)	Interaction mediated
Z boson	Z	1	0	91.1876 ± 0.0021	Weak interaction
Gluon	g	1	0	0	Strong interaction
Higgs boson	H^0	0	0	~125	Mass

2.1.2 Composite particles

Quarks are held together by the strong force to form composite particles. These composite particles are called hadrons. Hadrons are categorized into two groups. One group is the meson which is made of a quark and an anti-quark, the other group is the baryon which is made of three quarks. There is a quantum number called the baryon number for hadrons. It is defined as

$$B = \frac{1}{3}(n_q - n_{\bar{q}}) \tag{2.1}$$

where n_q is the number of quarks and $n_{\bar{q}}$ is the number of anti quarks. Another quantum number assigned to the hadrons is called the strangeness and is defined as

$$S = -(n_s - n_{\bar{s}}) \tag{2.2}$$

where n_s is the number of strange quarks and $n_{\bar{s}}$ is the number of strange anti-quarks. A third quantum number assigned to hadrons is the third component of isospin. It is defined as

$$I_3 = \frac{1}{2}[(n_u - n_{\bar{u}}) - (n_d - n_{\bar{d}})] \tag{2.3}$$

where n_u ($n_{\bar{u}}$) is the number of up quarks (anti-quarks) and n_d ($n_{\bar{d}}$) is the number of down quarks (anti-quarks).

Quarks (anti-quarks), which contribute to the quantum numbers of hadrons, are called valence quarks (anti-quarks). Valence quarks are different to sea quarks. Sea quarks are virtual quark anti-quark pairs ($q\bar{q}$) and they do not contribute to hadron quantum numbers.

Mesons

Mesons are made up of a valence quark and a valence anti-quark. They are bosons with integer spins. The baryon number for a meson is 0. As the quark and anti-quark carry oppositecolor charges, the meson has zero total color charge and thus it is a color singlet. Figure. 2.1 shows quark compositions of mesons with a spin-0. Mesons on a line, which is perpendicular to the Q-axis, share an equal electric charge. Mesons on a line, which is perpendicular to the I_3-axis, have an equal I_3. Mesons on a line, which is perpendicular to the S-axis, take an equal strangeness.

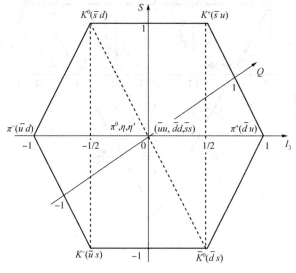

Figure 2.1   A meson nonet

Baryons

Baryons are made up of three valence quarks or three valence anti-quarks. They are fermions with half integer spins. Their baryon numbers are ± 1. As these three quarks (anti-quarks) carry different color charges: red, blue and green, the baryon has zero total color charge and thus it is a color singlet. Figure. 2.2(a) shows quark compositions for baryons with a spin-3/2. Figure. 2.2 (b) shows quark compositions for baryons with a spin-1/2.

2.1.3   Gauge theories

In particle physics, a Lagrangian is utilized to summarize the dynamics of particles and their interactions. QFT treats a particle as an excited state of an underlying physical field. The equations of motion for the underlying fields are determined by substituting the Lagrangian into the Euler-Lagrange equation for that field.

An important feature of the Lagrangian is that it is invariant under some transformations. If it is invariant under some global transformations, such as spatial translations or rotations, then there is a global symmetry for that Lagrangian. If it is invariant under some local transformations which depend on space-time coordinates, then there is a local symmetry for that Lagrangian.

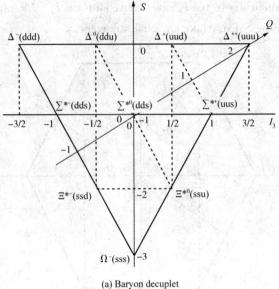

(a) Baryon decuplet

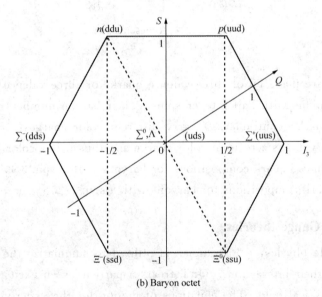

(b) Baryon octet

Figure 2.2 Baryon decuple and octet

The local transformations are called gauge transformations. The invariance of the Lagrangian under these local transformations is called gauge invariance. These local gauge transformations form a group, which is called a gauge symmetry group. One or several group generators are associated to that symmetry group. For each generator of the group, there is a corresponding vector field, which is called a gauge field. The corresponding particle to that field is called a gauge boson. If the generators of the group commute with each other, then that group is an abelian group. On the other hand, if they do not commute with each other, then that group is a non-abelian group. QED is a gauge theory with a symmetry group $U(1)$. As there is only one generator for $U(1)$, QED is an abelian gauge theory and there is only one gauge field. QCD is a gauge theory with a symmetry group $SU(3)$. As there are eight generators, which do not commute with each other, QCD is a non-abelian gauge theory and there are eight gauge fields. The standard model is a non-abelian gauge theory with a symmetry group $U(1) \times SU(2) \times SU(3)$. As there are a total of twelve generators for the standard model: one for $U(1)$, three for $SU(2)$ and eight for $SU(3)$, there are correspondingly twelve gauge bosons: one photon, three weak bosons and eight gluons.

2.1.4 Quantum electrodynamics

QED describes the interaction between charged spin-1/2 particles and photons. Here we will show that the QED Lagrangian, $\mathscr{L}_{QED}$, is invariant under a local gauge transformation of $U(1)$. $\mathscr{L}_{QED}$ is written as follows[2]:

$$\mathscr{L}_{QED} = \bar{\psi}(i\gamma^\mu \partial_\mu - m)\psi - \frac{1}{4}(F_{\mu\nu})^2 - e\bar{\psi}\gamma^\mu \psi A_\mu \qquad (2.4)$$

where the first term, $\bar{\psi}(i\gamma^\mu \partial_\mu - m)\psi$, is the Lagrangian for a free fermion of mass m, ψ is a bi-spinor field which describes the fermion field, $\bar{\psi} = \psi^\dagger \gamma^0$ is an adjoint field for ψ, γ^μ are Dirac matrices with $\mu = 0, 1, 2, 3$, $\gamma^\mu \partial_\mu$ is a contraction between the partial derivative ∂_μ and Dirac matrices γ^μ. The second term, $-\frac{1}{4}(F_{\mu\nu})^2$, is the Lagrangian for the electromagnetic field, A_μ is the electromagnetic vector potential, $F_{\mu\nu} = \partial_\mu A_\nu - \partial_\nu A_\mu$ is the electromagnetic field tensor. The third term, $-e\bar{\psi}\gamma^\mu \psi A_\mu$, is the Lagrangian describing the interaction between the fermion field and electromagnetic field, e is the electron charge. In order to show that the QED Lagrangian is invariant under $U(1)$, we rewrite it as the following one with the combination of the first term and the third term:

$$\mathscr{L}_{QED} = \bar{\psi}(i\gamma^\mu D_\mu - m)\psi - \frac{1}{4}(F_{\mu\nu})^2 \qquad (2.5)$$

where $\gamma^\mu D_\mu$ is a contraction between the gauge covariant derivative, $D_\mu \equiv \partial_\mu + ieA_\mu(x)$, and Dirac matrices γ^μ. With the local gauge transformation of $U(1)$,

$$\psi(x) \to e^{i\alpha(x)}\psi(x), A_\mu \to A_\mu - \frac{1}{e}\partial_\mu \alpha(x) \qquad (2.6)$$

where $\alpha(x)$ is a space-time dependent phase, the covariant derivative changes as follows:

$$D_\mu \psi(x) \to \left[\partial_\mu + ie\left(A_\mu - \frac{1}{e}\partial_\mu \alpha\right)\right] e^{i\alpha(x)} \psi(x)$$
$$= e^{i\alpha(x)} (\partial_\mu + ie A_\mu) \psi(x) = e^{i\alpha(x)} D_\mu \psi(x) \qquad (2.7)$$

As behaving in the same way as the transformation of $\psi(x)$, this covariant derivative transformation keeps the first term of the QED Lagrangian in Eq. (2.5) invariant. Another term, $\frac{1}{4}(F_{\mu\nu})^2$, is also invariant under the local gauge transformation. Thus the whole QED Lagrangian is invariant under this gauge transformation. As there is no mass term, $A^\mu A_\mu$, which is forbidden by gauge invariance, photons are massless.

2.1.5 Quantum chromodynamics

QCD describes the strong interaction between quarks and gluons. Here we will show that the QCD Lagrangian is invariant under a local gauge transformation of $SU(3)$. The QCD Lagrangian, $\mathscr{L}_{\text{QCD}}$, is written in a similar way as the QED Lagrangian in Eq. (2.5). It is described as follows[2]:

$$\mathscr{L}_{\text{QCD}} = \bar{\psi}_f (i\gamma^\mu D_\mu - m_f) \psi_f - \frac{1}{4}(F^a_{\mu\nu})^2 \qquad (2.8)$$

where ψ_f is the quark field. As there are three colors for quarks, this quark field is a triplet of the bi-spinor field. m_f is the mass of the quark with a flavour f. $\gamma^\mu D_\mu$ is a contraction between the gauge covariant derivative D_μ and Dirac matrices γ^μ. The gauge covariant derivative D_μ is defined as follows:

$$D_\mu \equiv \partial_\mu - ig A^a_\mu t^a \qquad (2.9)$$

where t^a are the generators of the $SU(3)$ group, $a = 1, 2, \cdots, 8$. This covariant derivative requires eight gluon vector fields A^a_μ, one for each generator of the $SU(3)$ group. g is the coupling constant between quarks and gluons. The gluon field tensor is similar to the electromagnetic field tensor and it is written as follows:

$$F^a_{\mu\nu} = \partial_\mu A^a_\nu - \partial_\nu A^a_\mu + g f^{abc} A^b_\mu A^c_\nu \qquad (2.10)$$

where f^{abc} are structure constants for $SU(3)$, f^{abc} determines the commutation relations between the generators of $SU(3)$ as follows:

$$[t^a, t^b] = i f^{abc} t^c \qquad (2.11)$$

The third term in Eq. (2.10) is new, it comes from the interaction between gluons[1]. The local gauge $SU(3)$ transformation is similar to the local gauge $U(1)$

---

[1] Gluons carry color charges. They can interact with each other. Photons do not carry electric charges, thus there is no such term in QED.

transformation in Eq. (2.6) and it is as follows:

$$\psi(x) \to e^{i\alpha^a(x) t^a} \psi(x), A_\mu^a t^a \to A_\mu^a + ig(\partial_\mu \alpha^a) t^a + i[\alpha^a t^a, A_\mu^b t^b] + \cdots \quad (2.12)$$

The generator in $U(1)$, **1**, is substituted with the generator in $SU(3)$, t^a. $\alpha(x)$ in Eq. (2.6) is replaced by $\alpha^a(x)$, one for each generator t^a. As t^a are not commutative with each other, the last term in the A^μ transformation law is new. In order to see that the QCD Lagrangian is invariant under the $SU(3)$ gauge transformation, we write down the infinitesimal form of this transformation:

$$\psi(x) \to (1 + i\alpha^a(x) t^a) \psi(x) \quad (2.13)$$

With this infinitesimal transformation, the covariant derivative changes as follows:

$$D_\mu \psi(x) \to (\partial_\mu - ig A_\mu^a t^a - i(\partial_\mu \alpha^a) t^a + g[\alpha^a t^a, A_\mu^b t^b])(1 + i\alpha^c(x) t^c) \psi(x)$$
$$= (1 + i\alpha^a(x) t^a)(\partial_\mu - ig A_\mu^a t^a) \psi(x)$$
$$= (1 + i\alpha^a(x) t^a) D_\mu \psi(x) \quad (2.14)$$

It is calculated up to terms of α^2. As this infinitesimal transformation for $D_\mu \psi(x)$ has the same form as the infinitesimal transformation for $\psi(x)$, it keeps the first term of the QCD Lagrangian invariant. The second term of the QCD Lagrangian, $-\frac{1}{4}(F_{\mu\nu}^a)^2$, is also invariant under the $SU(3)$ gauge transformation in Eq. $(2.12)^2$. Thus the whole QCD Lagrangian is invariant under this transformation. In order to see that gluons are massless, we insert Eq. (2.10) into the second term of the QCD Lagrangian. This term then is written as follows:

$$-\frac{1}{4}(\partial_\mu A_\nu^a - \partial_\nu A_\mu^a)(\partial^\mu A^{\nu a} - \partial^\nu A^{\mu a})$$

$$-\frac{g}{2} f^{abc}(\partial_\mu A_\nu^a - \partial_\nu A_\mu^a) A^{\mu b} A^{\nu c} - \frac{g^2}{4} f^{abe} f^{cde} A^{\mu a} A^{\nu b} A_\mu^c A_\nu^d \quad (2.15)$$

where the second term is an interaction term among three gluons, the third term is an interaction term among four gluons. There is no mass term $A^\mu A_\mu$ as it is forbidden by gauge invariance. Thus gluons are massless.

2.1.6 Asymptotic freedom

Asymptotic freedom is an important feature of QCD. It says that in the high energy region quarks interact weakly and cross-sections in deep inelastic processes can be calculated in a perturbative way. In the low energy region, quarks interact strongly with an attractive force and they are confined in hadrons.

Asymptotic freedom can be derived with the calculation of a β function. This β function tells us how the strong coupling constant, g, depends on the energy scale, Q, in the interaction[2]. It is written as follows:

$$\beta(g) = \frac{\partial g}{\partial \log(Q)} \tag{2.16}$$

If $\beta > 0$, then the coupling constant will increase when the energy scale increases. If $\beta < 0$, then the coupling constant will decrease when the energy scale increases and thus asymptotic freedom appears. The β function for QCD is

$$\beta(g) = -\frac{g^3}{(4\pi)^2}\left(11 - \frac{2}{3}n_f\right) \tag{2.17}$$

where n_f is the number of quark flavours. There are 6 quark flavours for QCD, thus β is negative and QCD is an asymptotically free theory.

In order to understand the asymptotic freedom of QCD, we firstly try to understand the screening effect of QED. When a charged particle travels through a space, it will emit and reabsorb photons. This will cause the vacuum around the charged particle to be polarized with virtual charged particle anti-particle pairs, such as $e^+ e^-$ pairs. If virtual particles carry opposite charges, they are attracted to the charged particle. On the other hand, if virtual particles carry like charges, they are repelled to that charged particle. Thus the effect charge of this particle varies with distance. When the distance between the view point and the charged particle is shorter and shorter, the effective of vacuum polarization gets weaker and weaker, and the effective charge increases. As a result, the electromagnetic coupling constant, $\alpha_e = e^2/4\pi$, gets stronger when the energy scale, Q, increases.

In QCD, there is a similar screening effect due to virtual quark anti-quark ($q\bar{q}$) pairs. These $q\bar{q}$ screens the colour charge of a central quark. However, there is another anti-screening effect due to virtual gluons. As gluons carry color charges, these virtual gluon pairs augment the color charge of the central quark. This anti-screening effect is 12 times larger than the screening effect[2]. Thus the net effect is to amplify the effective color charge at a large distance. The coupling constant of QCD grows larger at a larger distance due to this amplification. As a result, quarks can not be separated as it requires an infinite amount of energy to isolate them.

2.1.7 Spontaneous symmetry breaking

In order to understand the Higgs mechanism, firstly we need to know about spontaneous symmetry breaking (SSB). SSB is a model of symmetry breaking in a system. In that system, equations of motion or the system's Lagrangian obey certain symmetries, but these symmetries are broken in the lowest energy solutions of that system. A ϕ^4 theory is utilized to illustrate SSB. When the dimension of the scalar field, ϕ is N, $(N-1)$ massless bosons will appear through SSB. In order to understand this phenomenon, firstly we set the dimension of the scalar field to be one, and then we consider the case with N dimensions.

The Lagrangian for the ϕ^4 theory with one dimensional scalar field is as follows[2]:

$$\mathscr{L} = \frac{1}{2}(\partial_\mu \phi)^2 - \frac{1}{2} m^2 \phi^2 - \frac{\lambda}{4!}\phi^4 \qquad (2.18)$$

where $\lambda > 0$ is a dimensionless coupling constant. The kinetic energy is $E_{kin} = \frac{1}{2}(\partial_\mu \phi)^2$ and the potential is $V(\phi) = \frac{1}{2} m^2 \phi^2 + \frac{\lambda}{4!}\phi^4$. The Lagrangian has a symmetry, $\phi \rightarrow -\phi$. In order to show this symmetry is spontaneously broken around the lowest energy, we minimize the potential

$$\frac{\partial V(\phi)}{\partial \phi} = 0 \Rightarrow \phi\left(m^2 + \frac{\lambda}{3!}\phi^2\right) = 0 \qquad (2.19)$$

The potential is minimum at the following values of ϕ:

$$(\phi)_0 = \begin{cases} 0, \text{for } m^2 > 0 \\ \pm\sqrt{-\frac{6 m^2}{\lambda}} \neq 0, \text{for } m^2 < 0 \end{cases} \qquad (2.20)$$

where the first one is trivial with $m^2 > 0$ while the second one is non-trivial with $m^2 < 0$. The trivial case is not interesting and we only focus on the $m^2 < 0$ case. The non-trivial value is called the vacuum expectation value of ϕ and it is rewritten as follows:

$$(\phi)_0 = \pm v = \pm\sqrt{\frac{6}{\lambda}}\mu \qquad (2.21)$$

where m^2 is replaced by a negative parameter, $-\mu^2$. The potential for the non-trivial case is shown in Figure. 2.3.

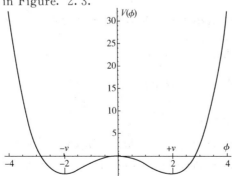

Figure 2.3 Potential for the ϕ^4 theory with the non-trivial vacuum expectation value. Here $m^2 = -4$, $\lambda = 6$

If the system is near the positive minima, then we rewrite ϕ in terms of v and a new shifted field σ as follows:

$$\phi(x) = v + \sigma(x) \qquad (2.22)$$

With the substitution of Eq. (2.22) into Eq. (2.18), the ϕ^4 Lagrangian is changed as follows:

$$\mathcal{L} = \frac{1}{2}(\partial_\mu \sigma)^2 - \frac{1}{2}(2\mu^2)\sigma^2 - \sqrt{\frac{\lambda}{6}}\mu\sigma^3 - \frac{\lambda}{4!}\sigma^4 + \text{const.} \qquad (2.23)$$

where the linear term of σ has vanished. The shifted scalar field, σ, now has a new mass of $\sqrt{2}\mu$. The vacuum expectation value, σ_0, now is 0. Due to the σ^3 term, the Lagrangian is not symmetric any more with a $\sigma \to -\sigma$ transformation.

In order to see the appearance of massless scalar bosons, we consider a scalar field with N dimensions, $\phi^i(x)$, $i = 1, 2, \cdots, N$. The Lagrangian for this scalar field is as follows[2]:

$$\mathcal{L} = \frac{1}{2}(\partial_\mu \phi^i)^2 - \frac{1}{2}m^2(\phi^i)^2 - \frac{\lambda}{4!}[(\phi^i)^2]^2 \qquad (2.24)$$

where $(\phi^i)^2 = (\phi^1)^2 + (\phi^2)^2 + \cdots + (\phi^N)^2$. This Lagrangian has a continuous symmetry, $\phi^i \to R^{ij}\phi^j$, where R^{ij} is a $N \times N$ orthogonal matrix. The group of this transformation is called $O(N)$. The potential, $V(\phi^i) = \frac{1}{2}m^2(\phi^i)^2 + \frac{\lambda}{4!}[(\phi^i)^2]^2$, is minimized in order to get the vacuum expectation value. The non-trivial expectation value is $(\phi^i)^2_0 = \frac{\mu^2}{\lambda}$ where $-\mu^2 = m^2$ is a negative parameter. Only the length of the vector ϕ^i_0 is determined by this expectation value, the direction of the vector is arbitrary. Here we let ϕ^i_0 point towards the Nth direction and write it as follows:

$$\phi^i_0 = (0, 0, \cdots, v) \qquad (2.25)$$

where $v = \frac{\mu}{\sqrt{\lambda}}$. With a transformation, $\phi^k \to \pi^k = \phi^k$, $\phi^N \to \sigma = \phi^N - v$, $k = 1, 2, \cdots, N-1$, the Lagrangian becomes

$$\mathcal{L} = \frac{1}{2}(\partial_\mu \pi^k)^2 + \frac{1}{2}(\partial_\mu \sigma)^2 - \frac{1}{2}(2\mu^2)\sigma^2 - \frac{\lambda}{4!}[(\pi^k)^2 + \sigma^2]^2$$

$$- \sqrt{\lambda}\mu\sigma^3 - \sqrt{\lambda}\mu(\pi^k)^2\sigma \qquad (2.26)$$

The σ field gets a new mass with $m_\sigma = \sqrt{2}\mu$ while a set of $(N-1)$ π fields are massless. The vacuum expectation values for π and σ now are 0. The new Lagrangian has a continuous symmetry, $\pi^k \to R^{kl}\pi^l$, where R^{kl} is a $(N-1) \times (N-1)$

orthogonal matrix. The massless π field is known as a Goldstone boson. The difference between the number of generators before and after the SSB will give the number of Goldstone bosons. Before SSB, there are $N(N-1)/2$ generators for the $O(N)$ group. After SSB, there are $(N-1)(N-2)/2$ generators for the $O(N-1)$ group. Thus there are $(N-1)$ Goldstone bosons. This result is known as the Goldstone's theorem.

2.1.8 Higg smechanism

Sections 2.1.4 and 2.1.5 show that in a local gauge symmetry a massless vector field is required for each generator of the symmetry. Section 2.1.7 illustrates that when a global symmetry is spontaneously broken, a Goldstone boson appears, with one for each generator of the spontaneously broken symmetry. In this section, the local gauge invariance and spontaneous symmetry breaking are combined together in a theory, known as the Higgs mechanism. Gauge bosons such as $W^\pm$ and Z acquire masses through spontaneous symmetry breaking of the gauge symmetry.

In order to describe the Higgs mechanism, a model with a $SU(2)$ gauge field coupled to a scalar field ϕ is utilized. ϕ is a doublet of two complex scalar fields:

$$\phi(x) = \begin{pmatrix} \phi_1(x) + i\phi_3(x) \\ \phi_2(x) + i\phi_4(x) \end{pmatrix} \qquad (2.27)$$

It is convenient to write ϕ as a doublet of two real scalar fields with a unitarity gauge[2]:

$$\phi(x) = \begin{pmatrix} \phi_1(x) \\ \phi_2(x) \end{pmatrix} \qquad (2.28)$$

Then the Lagrangian for this model is

$$\mathcal{L} = |D_\mu \phi|^2 - V(\phi) - \frac{1}{4}(F_{\mu\nu}^a)^2 \qquad (2.29)$$

where $V(\phi) = -\mu^2 \phi^\dagger \phi + \frac{\lambda}{2}(\phi^\dagger \phi)^2 = -\mu^2 \phi_i \phi_i + \frac{\lambda}{2}(\phi_i \phi_i)^2$ with $i = 1, 2$. $D_\mu \phi = (\partial_\mu - ig A_\mu^a \tau^a)\phi$, τ^a are the $SU(2)$ generators and $\tau^a = \sigma^a/2$, $a = 1, 2, 3$. g is the coupling constant between the gauge vector fields, A_μ^a, and ϕ and it is different to g in QCD, which is the coupling constant between quarks and gluons. $F_{\mu\nu}^a = \partial_\mu A_\nu^a - \partial_\nu A_\mu^a + g\epsilon^{abc} A_\mu^b A_\nu^c$, ϵ^{abc} are structure constants of $SU(2)$, $\epsilon^{123} = 1$. This Lagrangian has a $SU(2)$ gauge symmetry under the following transformation:

$$\phi(x) \to e^{i\alpha^a(x)\tau^a}\phi(x), A_\mu^a \tau^a \to A_\mu^a + \frac{1}{g}(\partial_\mu \alpha^a)\tau^a + i[\alpha^a \tau^a, A_\mu^b \tau^b] + \cdots$$

$$(2.30)$$

In order to see this $SU(2)$ gauge symmetry is spontaneously broken, $V(\phi)$ is minimized. The non-trivial vacuum expectation value of ϕ is

$$(\phi)_0 = \frac{1}{\sqrt{2}} \begin{pmatrix} 0 \\ v \end{pmatrix} \qquad (2.31)$$

where $v = \sqrt{\frac{2\mu^2}{\lambda}}$. A new shifted field is defined as follows:

$$\phi' = \phi - (\phi)_0 \qquad (2.32)$$

The Lagrangian in terms of the new shifted field is no longer symmetric under this $SU(2)$ transformation. In order to see the result of spontaneous symmetry breaking, we rewrite the kinetic energy term as follows:

$$|D_\mu \phi|^2 = \frac{1}{2} g^2 (0 \quad v) \tau^a \tau^b \begin{pmatrix} 0 \\ v \end{pmatrix} A_\mu^a A^{\mu b} + \cdots \qquad (2.33)$$

The first term contains the gauge boson mass term. This mass term can be written as

$$\Delta \mathscr{L} = \frac{g^2 v^2}{8} A_\mu^a A^{a\mu} \qquad (2.34)$$

As a result, the spontaneous gauge symmetry breaking gives each of the three gauge bosons a mass,

$$m_A = \frac{gv}{2} \qquad (2.35)$$

2.1.9 Electroweak theory

Now let's turn our attention to the electroweak interaction. There are four gauge bosons in the electroweak theory: $W^\pm$, Z and γ. $W^\pm$ and Z are massive while γ is massless. In section 2.1.8 we see that the model with a $SU(2)$ gauge field coupled to a scalar field gives three massive gauge bosons. In order to get another massless gauge boson, we introduce another $U(1)$ gauge symmetry to that model. As a result, the electroweak theory is a $SU(2) \times U(1)$ gauge theory.

In this section, firstly we will show the interactions between the gauge bosons and the scalar field. This explains how these gauge bosons acquire masses through the Higgs mechanism. Then we will describe the couplings between the gauge bosons and fermions. Finally we will illustrate the interactions between the fermions and the scalar field. This interprets how the fermions acquire masses.

Electroweak gauge bosons masses

A charge $+1/2$ is assigned to the scalar field under the $U(1)$ gauge symmetry. The covariant derivative on the scalar field ϕ now is[2]

$$D_\mu \phi = \left(\partial_\mu - ig A_\mu^a \tau^a - i \frac{1}{2} g' B_\mu \right)\phi \qquad (2.36)$$

where A_μ^a are the $SU(2)$ gauge bosons while B_μ is the $U(1)$ gauge boson, g' is coupling constant between B_μ and ϕ. The kinetic energy term for the scalar field ϕ can be rewritten in terms of the new field defined in Eq. (2.32) as

$$|D_\mu \phi|^2 = \frac{1}{2}(0 \ \ v)\left(g A_\mu^a \tau^a + \frac{1}{2} g' B_\mu \right)\left(g A^{b\mu} \tau^b + \frac{1}{2} g' B^\mu \right)\binom{0}{v} A_\mu^a A^{b\mu} + \cdots \qquad (2.37)$$

The first term contains the mass term for the gauge bosons A_μ^a and B_μ. This mass term can be rewritten as

$$\Delta \mathscr{L} = \frac{1}{2}\frac{v^2}{4}[g^2 (A_\mu^1)^2 + g^2 (A_\mu^2)^2 + (-g A_\mu^3 + g' B_\mu)^2] \qquad (2.38)$$

when the matrix product in Eq. (2.37) is evaluated in detail. It indicates that three gauge bosons are massive while another gauge boson is massless. The first two massive bosons are $W^\pm$, $W^\pm = \frac{1}{\sqrt{2}}(A_\mu^1 \mp i A_\mu^2)$, their masses are $m_{W^\pm} = g \frac{v}{2}$. The third massive boson is Z, $Z_\mu = \frac{1}{\sqrt{g^2 + g'^2}}(g A_\mu^3 - g' B_\mu)$, its mass is $m_Z = \sqrt{g^2 + g'^2}\,\frac{v}{2}$. The fourth massless boson is A_μ, $A_\mu = \frac{1}{\sqrt{g^2 + g'^2}}(g A_\mu^3 + g' B_\mu)$. A_μ^3 and B_μ are coalesced into the mass eigenstate bosons A_μ and Z_μ in the following way:

$$\begin{pmatrix} Z_\mu \\ A_\mu \end{pmatrix} = \begin{pmatrix} \cos\theta_W & -\sin\theta_W \\ \sin\theta_W & \cos\theta_W \end{pmatrix}\begin{pmatrix} A_\mu^3 \\ B_\mu \end{pmatrix} \qquad (2.39)$$

where θ_W is the weak mixing angle, $\cos\theta_W = \frac{g}{\sqrt{g^2 + g'^2}}$, $\cos\theta_W = \frac{g'}{\sqrt{g^2 + g'^2}}$. The relation between the masses of W and Z is $m_W = m_Z \cos\theta_W$.

Gauge bosons coupled to fermions

A hyper-charge, Y, is assigned to the fermion field under the $U(1)$ gauge symmetry. The covariant derivative on the fermion field[2],

$$D_\mu = \partial_\mu - ig A_\mu^a T^a - i g' Y B_\mu \qquad (2.40)$$

can be written in terms of the mass eigenstate fields as

$$D_\mu = \partial_\mu - i \frac{e}{\sqrt{2}\sin\theta_W}(W_\mu^+ T^+ + W_\mu^- T^-) - i \frac{e}{\sin\theta_W \cos\theta_W} Z_\mu (T^3 - \sin^2\theta_W Q) - ie A_\mu Q \qquad (2.41)$$

where e is the electron charge and $e = \dfrac{g g'}{\sqrt{g^2 + g'^2}}$, Q is the electric charge quantum number and $Q = T^3 + Y$, $T^{\pm} = (T^1 + i T^2) = \dfrac{1}{2}(\sigma^1 \pm i \sigma^2) = \sigma^{\pm}$, $T^3 = \dfrac{1}{2}\sigma^3$, T^3 refers to the third component of the weak isospin. It is obvious that the couplings of all electroweak bosons to fermion fields are described by two parameters: the electron charge e and the weak mixing angle θ_W.

A bi-spinor fermion field can be decomposed as

$$\psi = \begin{pmatrix} \psi_L \\ \psi_R \end{pmatrix} \tag{2.42}$$

where ψ_L and ψ_R are left-handed❶ and right-handed fermion fields. ψ_L and ψ_L can be rewritten as

$$\begin{cases} \psi_L = P_L \psi = \dfrac{1}{2}(1 - \gamma^5)\psi \\ \psi_R = P_R \psi = \dfrac{1}{2}(1 + \gamma^5)\psi \end{cases} \tag{2.43}$$

where P_L and P_R are left-handed and right-handed projection operators, $\gamma^5 = i \gamma^0 \gamma^1 \gamma^2 \gamma^3$, $\bar{\psi}_L = \bar{\psi} P_R$, $\bar{\psi}_R = \bar{\psi} P_L$, $P_L^2 = P_L$, $P_R^2 = P_R$, $P_L \cdot P_R = 0$. Left-handed and right-handed currents are written as

$$\begin{cases} j_L^\mu = \bar{\psi}_L \gamma^\mu \psi_L = \bar{\psi} \gamma^\mu \left(\dfrac{1 - \gamma^5}{2}\right) \psi \\ j_R^\mu = \bar{\psi}_R \gamma^\mu \psi_R = \bar{\psi} \gamma^\mu \left(\dfrac{1 + \gamma^5}{2}\right) \psi \end{cases} \tag{2.44}$$

where $\bar{\psi} \gamma^\mu \psi$ is called a vector current, $\bar{\psi} \gamma^\mu \gamma^5 \psi$ is called an axial vector current. Thus the left-handed (right-handed) current is a mixture of the vector and axial vector currents.

In nature, the W boson only couples to left-handed fermions while the Z boson couples to both left-handed and right-handed fermions. Left-handed fermions transform as a doublet of $SU(2)$. Their first generation is

$$E_L = \begin{pmatrix} \nu_e \\ e \end{pmatrix}_L, \quad Q_L = \begin{pmatrix} u \\ d \end{pmatrix}_L \tag{2.45}$$

Right-handed fermions transform as a singlet of $SU(2)$. Their first generation is

---

❶ Here handedness is different to helicity. Helicity is the projection of a fermion's spin along its momentum. A fermion with a left-handed helicity has a momentum whose direction is opposite to its spin. A fermion with a right-handed helicity has a momentum whose direction is the same as its spin. Helicity is the same as handedness if the fermion moves at the speed of light.

e_R^-, u_R and d_R. T^3 is assigned to $\pm\frac{1}{2}$ for the up-type and down-type left-handed fermions respectively. For the right-handed fermions, T^3 is assigned to 0. The assignments of T^3, Y and Q for the left-handed and right-handed leptons and quarks are listed in Table 2.4.

Table 2.4 The assignments of T^3, Y and Q for leptons and quarks

Particles	T^3	Y	Q
ν_{eL}	$+\frac{1}{2}$	$-\frac{1}{2}$	0
e_L	$-\frac{1}{2}$	$-\frac{1}{2}$	-1
u_L	$+\frac{1}{2}$	$+\frac{1}{6}$	$+\frac{2}{3}$
d_L	$-\frac{1}{2}$	$+\frac{1}{6}$	$-\frac{1}{3}$
u_R	0	$+\frac{2}{3}$	$+\frac{2}{3}$
d_R	0	$-\frac{1}{3}$	$-\frac{1}{3}$
e_R	0	-1	-1

Now we can write down the Lagrangian for the interaction between the electroweak gauge bosons and fermions. It is as follows:

$$\mathcal{L} = \bar{E}_L(i\gamma^\mu D_\mu)E_L + \bar{e}_R(i\gamma^\mu D_\mu)e_R + \bar{Q}_L(i\gamma^\mu D_\mu)Q_L$$
$$+ \bar{u}_R(i\gamma^\mu D_\mu)u_R + \bar{d}_R(i\gamma^\mu D_\mu)d_R \qquad (2.46)$$

This Lagrangian can be rewritten in terms of the gauge boson mass eigenstates as follows:

$$\mathcal{L} = \bar{E}_L(i\gamma^\mu \partial_\mu)E_L + \bar{e}_R(i\gamma^\mu \partial_\mu)e_R + \bar{Q}_L(i\gamma^\mu \partial_\mu)Q_L + \bar{u}_R(i\gamma^\mu \partial_\mu)u_R$$
$$+ \bar{d}_R(i\gamma^\mu \partial_\mu)d_R + g(W_\mu^+ J_W^{\mu+} + W_\mu^- J_W^{\mu-} + Z_\mu J_Z^\mu) + eA_\mu J_{EM}^\mu \qquad (2.47)$$

where $J_W^{\mu+}$ ($J_W^{\mu-}$) is the charged weak current coupled to W_μ^+ (W_μ^-), $J_W^{\mu+} = \frac{1}{\sqrt{2}}(\bar{\nu}_L\gamma^\mu e_L + \bar{u}_L\gamma^\mu d_L)$, $J_W^{\mu-} = \frac{1}{\sqrt{2}}(\bar{e}_L\gamma^\mu \nu_L + \bar{d}_L\gamma^\mu u_L)$, J_Z^μ is the neutral weak current coupled to Z_μ, $J_Z^\mu = \frac{1}{\cos\theta_W}\left[\bar{\nu}_L\gamma^\mu \frac{1}{2}\nu_L + \cdots\right]$, J_{EM}^μ is the electromagnetic current couple to A_μ, $J_{EM}^\mu = \bar{e}\gamma^\mu(-1)e + \cdots$ ².

Fermion masses

Since ψ_L and ψ_R have different hyper-charges, the gauge transformations for

ψ_L and ψ_R are different. As a result, the fermion mass term, which is proportional to $\bar{\psi}\psi = \bar{\psi}_L \psi_R + \bar{\psi}_R \psi_L$, violates the gauge invariance, and it is forbidden in the electroweak Lagrangian. However fermions can acquire masses through SSB when they couple to the scalar field. The interaction between the fermion fields and scalar field is described by the Yukawa theory[1]. The Lagrangian for this interaction is given by[2]

$$\mathcal{L}_{\text{Yukawa}} = -\lambda_e \bar{E}_L \phi e_R - \lambda_d \bar{Q}_L \phi d_R - \lambda_u \epsilon^{ab} \bar{Q}_{La} \phi_b u_R + h.c. \qquad (2.48)$$

where λ_e, λ_d and λ_u are coupling constants, $h.c.$ is the hermitian conjugate of the first three terms. This Lagrangian can be rewritten in terms of the shifted scalar field defined in Eq. (2.32) as follows:

$$\mathcal{L}_{\text{Yukawa}} = -\frac{1}{\sqrt{2}} \lambda_e v \bar{e}_L e_R - \frac{1}{\sqrt{2}} \lambda_d v \bar{d}_L d_R - \frac{1}{\sqrt{2}} \lambda_u v \bar{u}_L u_R + h.c. + \cdots \qquad (2.49)$$

The first three terms indicate that the electron, d quark and u quark acquire masses. Their masses are

$$m_e = \frac{1}{\sqrt{2}} \lambda_e v, \quad m_d = \frac{1}{\sqrt{2}} \lambda_d v, \quad m_u = \frac{1}{\sqrt{2}} \lambda_u v \qquad (2.50)$$

2.1.10 Higgs boson

In this section, we will show the strength of the couplings between the scalar field and gauge bosons as well as fermions. In section 2.1.9, we have seen that the scalar field plays an important role for $W^\pm$, Z and fermions getting masses. Now let's have a look at its properties. The scalar field ϕ can be rewritten as follows[2]:

$$\phi(x) = U(x) \frac{1}{\sqrt{2}} \begin{pmatrix} 0 \\ v + h(x) \end{pmatrix} \qquad (2.51)$$

where $h(x)$ is a real-value scalar field with $h(x)_0 = 0$, $\frac{v}{\sqrt{2}}$ is the non-trivial vacuum expectation value of ϕ. $U(x)^\dagger$ transforms a complex-valued two-component spinor into a real-valued two-component spinor.

The Lagrangian for the scalar field ϕ is

$$\mathcal{L} = |D_\mu \phi|^2 + \mu^2 \phi^\dagger \phi - \frac{\lambda}{2} (\phi^\dagger \phi)^2 \qquad (2.52)$$

As shown in Eq. (2.31), the non-trivial expectation value for ϕ is

---

[1] The Yukawa theory describes the interaction between a scalar field and a fermion field. The interaction term is proportional to $g \bar{\psi} \phi \psi$, where g is a coupling constant.

$$(\phi)_0 = \frac{1}{\sqrt{2}}\begin{pmatrix}0\\v\end{pmatrix} \qquad (2.53)$$

where $v = \sqrt{\frac{2\mu^2}{\lambda}}$. Plugging ϕ into the scalar field Lagrangian, we can get

$$\mathscr{L} = -\frac{1}{2}m_h^2 h^2 - \frac{\sqrt{\lambda}}{2}m_h h^3 - \frac{1}{8}\lambda h^4 + \cdots \qquad (2.54)$$

The first term indicates that the scalar field $h(x)$ gets a mass with $m_h = \sqrt{2}\mu$. $h(x)$ is known as the Higgs boson. Its mass can be rewritten in terms of the vacuum expectation value, v, and the coupling constant, λ, as follows:

$$m_h = \sqrt{\lambda}\, v \qquad (2.55)$$

It not only depends on the vacuum expectation value, but also depends on the coupling constant between the scalar fields, ϕ.

Now let's have a look at the coupling between the Higgs bosons and W or Z. Plugging ϕ in Eq. (2.51) into the covariant derivative in Eq. (2.36), we can obtain

$$\Delta\mathscr{L} = \frac{1}{2}(\partial_\mu h)^2 + \left[m_W^2 W^{\mu+} W_\mu^- + \frac{1}{2}m_Z^2 Z^\mu Z_\mu\right]\cdot\left(1+\frac{h}{v}\right)^2 \qquad (2.56)$$

The strength of the coupling between W bosons and the Higgs boson is $2\frac{m_W^2}{v}$. The strength of the coupling between Z and the Higgs boson is $2\frac{m_Z^2}{v}$.

Then let's have a look at the coupling between the Higgs boson and fermions. Plugging ϕ into the Lagrangian of the Yukawa theory in Eq. (2.46), we can get

$$\Delta\mathscr{L} = -m_f \bar{f} f \left(1+\frac{h}{v}\right) \qquad (2.57)$$

The strength of the coupling between fermions and the Higgs boson is $\frac{m_f}{v}$.

2.1.11 Quark mixing and CP violation

Quark mixing

A quark can change its flavor in one generation to another flavor in the same generation via its interaction with the W boson. For example, a d quark transitions to a u quark via the emission of a W^- boson. However, the quark can also change its flavor in one generation to another flavor in another generation. For example, a d quark transitions to a c quark via the emission of a W^- boson. The second example is explained by a mechanism called quark mixing. It says the weak quark eigenstates (d',s',b') are related to the physical mass eigenstates (d,s,b) by a matrix which is called the Cabibbo-Kobayashi-Maskawa (CKM) matrix, V. For the case of two generations, (d,s), V is a 2×2 unitary matrix which is written as follows[2]:

$$V = \begin{pmatrix} \cos\theta_c \, e^{i\alpha} & \sin\theta_c \, e^{i\beta} \\ -\sin\theta_c \, e^{i(\alpha+\gamma)} & -\cos\theta_c \, e^{i(\beta+\gamma)} \end{pmatrix} \quad (2.58)$$

There are four parameters: θ_c is a rotation angle which is called the Cabibbo angle; α, β and γ are phases which can be removed by the a global phase rotation on the quark fields. After removing the phases, we can rewrite the CKM matrix as follows:

$$V = \begin{pmatrix} \cos\theta_c & \sin\theta_c \\ -\sin\theta_c & -\cos\theta_c \end{pmatrix} \quad (2.59)$$

Thus the weak quark eigenstates (d',s') can be written in terms of the physical mass eigenstates (d,s) as follows:

$$\begin{cases} d' = \cos\theta_c d + \sin\theta_c s \\ s' = -\sin\theta_c d + \cos\theta_c s \end{cases} \quad (2.60)$$

It indicates that the probability for the transition of a $d(s)$ quark to a u quark, V_{ud}^2 (V_{us}^2), is $\cos^2\theta_c$ ($\sin^2\theta_c$), the probability for the transition of a c quark to a $d(s)$ quark, V_{cd}^2 (V_{cs}^2), is $\cos^2\theta_c$ ($\sin^2\theta_c$).

CP violation

Now let's turn to the case of three quark generations. There are 18 parameters for the CKM matrix, 9 of them are constrained by the unitarity of the CKM matrix, $V^\dagger V = 1$. For the other 9 parameters, three of them are rotation angles, the other six are phases. 5 of these phases can be removed by a global phase rotation on the quark fields. The last one is an overall phase and it is important for CP violation. CP symmetry is a product of the charge conjugation (C) symmetry and the parity (P) symmetry. The charge conjugation transforms a particle to its anti-particle and reverses the particle's quantum numbers such as the electrical charge. The parity transformation changes the sign of a particle's spatial coordinate and reverses the handedness of that particle. A left-handed electron is transformed into a right-handed positron under a CP transformation. CP symmetry is conserved in strong and electromagnetic interactions, but violated in weak interactions.

Let's take the inter-conversion of $\bar{K}^0 (s\bar{d})$ and $K^0 (d\bar{s})$ as an example to explain CP violation briefly[19]. The amplitude for $\bar{K}^0 \rightarrow K^0$ is M, $M = |M|e^{i\theta}$, and for the corresponding anti-particle process, $K^0 \rightarrow \bar{K}^0$, it is $\widetilde{M}$. Before CP violation, these two amplitudes should be the same, thus we write $\widetilde{M} = |M|e^{i\theta}$. If the overall phase from the CKM matrix, $e^{i\phi}$, is considered for the two processes, then $M = |M|e^{i\theta}e^{i\phi}$, $\widetilde{M} = |M|e^{i\theta}e^{-i\phi}$. So far there is no difference between the squared amplitudes of $|M|^2$ and $|\widetilde{M}|^2$. However, there are two processes contributing to $\bar{K}^0 \to K^0$. Their Feynman diagrams are shown in Figure. 2.4(a) and Figure. 2.4(b). The interference term between these two processes will contribute to CP violation. This is illustrated as follows[20]:

$$\begin{cases} M = |M_1|e^{i\theta_1}e^{i\phi_1} + |M_2|e^{i\theta_2}e^{i\phi_2} \\ \widetilde{M} = |M_1|e^{i\theta_1}e^{-i\phi_1} + |M_2|e^{i\theta_2}e^{-i\phi_2} \end{cases} \quad (2.61)$$

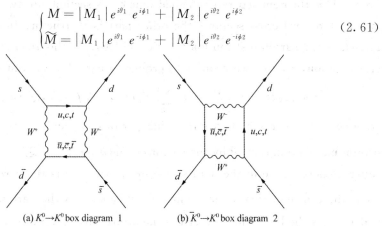

(a) $K^0 \to \bar{K}^0$ box diagram 1   (b) $\bar{K}^0 \to K^0$ box diagram 2

Figure 2.4 Feynman box diagrams for the $\bar{K}^0$-K^0 process

where $|M_1|e^{i\theta_1}e^{i\phi_1}$ is the amplitude for the left diagram while $|M_2|e^{i\theta_2}e^{i\phi_2}$ is the amplitude for the right diagram. After a detailed calculation, we can get

$$|M|^2 - |\widetilde{M}|^2 = -4|M_1||M_2|\sin(\theta_1 - \theta_2)\sin(\phi_1 - \phi_2) \quad (2.62)$$

The difference between the two amplitude squares is not 0. This result is an indicator of CP violation.

Although CP symmetry can be broken in weak interactions, it is believed that CPT is a perfect symmetry in the nature world. Here T is the time reversal transformation. If CP symmetry is violated, then the time reversal symmetry must be violated equally in the opposite direction of CP violation. At present, it is not observed that CPT is violated.

2.2 W Boson production at LHCb

This section is organized as follows. Firstly a general method utilized to calculate a given process cross-section is illustrated. Then this method is applied to the $W \to \mu\nu_\mu$ process. A μ^+ is produced in the $W^+ \to \mu^+\nu_\mu$ process, while a μ^- is produced in the $W^- \to \mu^- \bar{\nu}_\mu$ process. In order to understand this muon charge asymmetry and ratio, we derive the muon transverse momentum as well as angular distributions. Finally, we show the motivations to measure the cross-section for the $W \to \mu \nu_\mu$ process at LHCb.

2.2.1 QCD factorization theorem

The hadronic cross-section can be calculated from the QCD factorization theorem. This theorem separates the hadronic cross-section into two parts: one part is a parton level cross-section; the other part is a structure function, which is known as the parton distribution function (PDF). In this theorem, the hadronic cross-section, σ, for two protons to produce a particle, X, can be expressed as

$$\sigma_{pp \to X}(Q^2) = \int d x_a d x_b \sum_{a,b} f_a(x_a, Q^2) f_b(x_b, Q^2) \hat{\sigma}_{ab \to X} \quad (2.63)$$

where a and b represent two interacting partons, x_a and x_b are fractions of the proton momentum carried by the partons a and b, $x_{a(b)} = Q/\sqrt{s} \cdot e^{\pm y}$. Q^2 is the squared energy scale in the hard scattering process. y is the parton rapidity and $\sqrt{s}$ is the centre-of-mass energy in the collision. $\hat{\sigma}$ is the partonic cross-section and $f_{a(b)}$ is the PDF which describes the probability of a proton containing a parton $a(b)$ with the momentum fraction $x_{a(b)}$.

The parton level cross-section can be calculated in the QCD perturbative theory due to the fact that at a very high energy quarks in the proton are asymptotically free. The PDF can not be determined within the QCD perturbative theory since it is dependent on soft processes which determine the structure of the proton as a bound state of quarks and gluons. However, the PDF can be measured with a global fit to experimental data and it is assumed to be universal for different physical processes at the same Q^2 region. For different Q^2 regions the PDF will change slowly with Q^2 and this change is described by DGLAP evolution[21]. Figure. 2.5 shows the next-to-leading-order (NLO) parton distribution functions for quarks, anti-quarks and gluons in the proton at $Q^2 = 10$ GeV^2[22]. It is found that at high x regions u and d quarks tend to carry a large fraction of the proton's momentum while their anti-quarks and gluons tend to carry small fractions. The parton distribution functions are available from several groups worldwide. The major PDF sets are MSTW08[22], ABKM09[23], JR09[24], HERA1.5[25], NNPDF2.1[26], and CTEQ6.6[27].

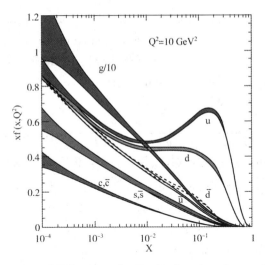

Figure 2.5 NLO parton distribution functions for quarks, anti-quarks and gluons in the proton at $Q^2 = 10$ GeV^2. These distributions are obtained by the MSTW collaboration. Taken from Ref.[22]

2.2.2 $W \to \mu \nu_\mu$ cross-section

Now let's turn to the $W \to \mu \nu_\mu$ cross-section calculation in pp collisions at $\sqrt{s}$ = 7 TeV. A down-type quark (q_i) from one proton (P_1) will annihilate with another up-type anti-quark ($\bar{q}_j$) in another proton (P_2) to produce an on-shell W^- boson. The W^- boson then will decay into a muon (μ^-) and a muon anti-neutrino ($\bar{\nu}_\mu$). The fraction of the proton's momentum carried by the quark (anti-quark) q_i ($\bar{q}_j$) is x_1 (x_2). Figure. 2.6 shows the Feynman diagram for the $W^- \to \mu^- \bar{\nu}_\mu$ process in the pp collision.

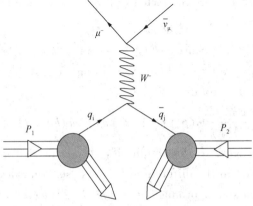

Figure 2.6 Feynman diagram for the $W \to \mu \nu_\mu$ process in the pp collision

The total hadronic cross-section for the $W^- \to \mu^- \bar{\nu}_\mu$ process is derived with the following procedures. In the first step, a partonic cross-section for the $W^- \to \mu^- \bar{\nu}_\mu$ process is calculated. For a process in which the incoming particles are A and B and the final particles are C and D, its cross-section in the center-of-mass (CM) frame is as follows[2]:

$$\left(\frac{d\sigma}{d\Omega}\right)_{cm} = \frac{1}{2E_A 2E_B |v_A - v_B|} \frac{|\vec{p}_C|}{(2\pi)^2 4 E_{cm}} |\mathcal{M}(A,B \to C,D)|^2 \quad (2.64)$$

where E_A (E_B) is the energy of A(B), $|v_A - v_B|$ is the relative velocity of the incoming particles A and B, $\vec{p}_C$ is the momentum of C, E_{cm} is the CM energy, $E_{cm} = E_A + E_B$, $\mathcal{M}(A,B \to C,D)$ is the amplitude for the process $A,B \to C,D$. In the second step, we integrate this differential cross-section over $d\Omega$ to get the total partonic cross-section. Finally, inserting this partonic cross-section into Eq. (2.63), we get the hadronic cross-section for the $W^- \to \mu^- \bar{\nu}_\mu$ process.

Firstly, let's try to derive the partonic $W^- \to \mu^- \bar{\nu}_\mu$ cross-section. The momentum carried by q_i ($\bar{q}_j$) is p (p'), the momentum carried by μ^- ($\bar{\nu}_\mu$) is k (k'), and the momentum carried by the W^- boson is q, where $q = p + p'$. We write down the amplitude for the $W^- \to \mu^- \bar{\nu}_\mu$ process in the pp collision with the application of Feynman rules for electroweak interactions[7] as follows:

$$i\mathcal{M} = \bar{v}(p') \frac{-ig_W \gamma^\mu (1-\gamma^5)}{2\sqrt{2}} u(p) \frac{-i(g_{\mu\nu} - q_\mu q_\nu / M_W^2)}{q^2 - M_W^2} \times \bar{u}(k) \frac{-ig_W \gamma^\nu (1-\gamma^5)}{2\sqrt{2}} v(k')$$

(2.65)

where M_W is the W boson mass, g_W is the strength of the coupling between the W boson and leptons. As we are only interested in the region where $q^2 \approx M_W^2$, the masses of quarks and leptons can be ignored. Let's pay attention to the second term of the W boson propagator. When q_ν are contracted with γ^ν, it gives the following term:

$$\bar{u}(k) \gamma^\nu q_\nu (1-\gamma^5) v(k') \quad (2.66)$$

where $\gamma^\nu q_\nu = \gamma^\nu k_\nu + \gamma^\nu k'_\nu$. Since $\bar{u}(k)\gamma^\nu k_\nu = 0$ and $\gamma^\nu k'_\nu v(k') = 0$, the second term of the W boson propagator contributes to nothing. Thus, we can rewrite the amplitude in Eq. (2.65) as follows:

$$i\mathcal{M} = \frac{ig_W^2}{8(q^2 - M_W^2)} \bar{v}(p') \gamma^\mu (1-\gamma^5) u(p) \bar{u}(k) \gamma_\mu (1-\gamma^5) v(k') \quad (2.67)$$

A complex conjugate to the amplitude, $\mathcal{M}^*$, is needed in order to calculate $\mathcal{M}^2$, which is essential to the differential W^- cross-section calculation. The conjugated term for the bi-spinor product, $\bar{v}(p')\gamma^\mu(1-\gamma^5)u(p)$, is

$$(\bar{v}(p')\gamma^\mu(1-\gamma^5)u(p))^\dagger = \bar{u}(p)\gamma^\mu(1-\gamma^5)v(p') \quad (2.68)$$

where we have applied the identities $(\gamma^5)^\dagger = \gamma^5$ and $\{\gamma^5, \gamma^\mu\} = 0$. The squared matrix element then is written as follows:

$$\mathcal{M}^2 = \frac{g_W^4}{64(q^2 - M_W^2)^2} (\bar{v}(p')\gamma^\mu (1-\gamma^5) u(p) \bar{u}(p) \gamma^\nu (1-\gamma^5) v(p'))$$
$$\times (\bar{u}(k) \gamma_\mu (1-\gamma^5) v(k') \bar{v}(k') \gamma_\nu (1-\gamma^5) u(k)) \qquad (2.69)$$

When averaging over the initial colors and spins of quarks and summing over the spins of final leptons, we rewrite the squared matrix elements as

$$\frac{1}{4} \times \frac{1}{3} \sum_{spin} \mathcal{M}^2 = \frac{1}{12} \cdot \frac{g_W^4}{64(q^2 - M_W^2)^2} \times tr(\gamma^\lambda p'_\lambda \gamma^\mu (1-\gamma^5) \gamma^\sigma p_\sigma \gamma^\nu (1-\gamma^5))$$
$$\times tr(\gamma^\lambda p'_\lambda \gamma^\mu (1-\gamma^5) \gamma^\sigma p_\sigma \gamma^\nu (1-\gamma^5)) \qquad (2.70)$$

with the application of identities $\sum_s u^s(p) \bar{u}^s(p) = \gamma^\lambda p_\lambda + m$ and $\sum_s v^s(p) \bar{v}^s(p) = \gamma^\lambda p_\lambda - m$. We denote $tr(\gamma^\lambda p'_\lambda \gamma^\mu (1-\gamma^5) \gamma^\sigma p_\sigma \gamma^\nu (1-\gamma^5))$ as $A^{\mu\nu}$ and $tr(\gamma^\lambda p'_\lambda \gamma^\mu (1-\gamma^5) \gamma^\sigma p_\sigma \gamma^\nu (1-\gamma^5))$ as $B_{\mu\nu}$. In order to calculate $A^{\mu\nu}$ and $B_{\mu\nu}$, we first review the trace theorems as follows[2]:

$$\begin{cases} tr(\gamma^\mu \gamma^\nu \gamma^\rho \gamma^\sigma) = 4(g^{\mu\nu} g^{\rho\sigma} - g^{\mu\rho} g^{\nu\sigma} + g^{\mu\sigma} g^{\nu\rho}) \\ tr(\gamma^\mu \gamma^\nu \gamma^\rho \gamma^\sigma \gamma^5) = -4i \varepsilon^{\mu\nu\rho\sigma} \end{cases} \qquad (2.71)$$

where $\varepsilon^{\mu\nu\rho\sigma}$ is a Levi-Civita symbol in four dimensions, $\varepsilon^{0123} = +1$, $g^{\mu\nu}$ is a metric tensor, $g^{\mu\nu} g_{\mu\nu} = 4$. With the application of Eq. (2.71), $A^{\mu\nu}$ and $B_{\mu\nu}$ are rewritten as follows:

$$\begin{cases} A^{\mu\nu} = 8[(p'^\mu p^\nu - p' \cdot p \, g^{\mu\nu} + p'^\nu p^\mu) + i \varepsilon^{\rho\mu\sigma\nu} p'_\rho p_\sigma] \\ B_{\mu\nu} = 8[(k_\mu k'_\nu - k \cdot k' \, g_{\mu\nu} + k_\nu k'_\mu) + i \varepsilon_{\lambda\mu\tau\nu} k^\lambda k'^\tau] \end{cases} \qquad (2.72)$$

Note that in $A^{\mu\nu}$ and $B_{\mu\nu}$ the first term is symmetric while the second term is antisymmetric in $\mu \leftrightarrow \nu$. The product of $A^{\mu\nu}$ and $B_{\mu\nu}$ is

$$A^{\mu\nu} B_{\mu\nu} = 64((p'^\mu p^\nu - p' \cdot p \, g^{\mu\nu} + p'^\nu p^\mu) \cdot (k_\mu k'_\nu - k \cdot k' \, g_{\mu\nu} + k_\nu k'_\mu))$$
$$- 64 \varepsilon^{\rho\sigma\mu\nu} \varepsilon_{\lambda\tau\mu\nu} p'_\rho p_\sigma k^\lambda k'^\tau \qquad (2.73)$$

With the substitution of $\varepsilon^{\rho\sigma\mu\nu} \varepsilon_{\lambda\tau\mu\nu} = -2(\delta^\rho_\lambda \delta^\sigma_\tau - \delta^\rho_\tau \delta^\sigma_\lambda)$, $A^{\mu\nu} B_{\mu\nu}$ is rewritten as follows:

$$A^{\mu\nu} B_{\mu\nu} = 128[(p' \cdot k)(p \cdot k') + (p' \cdot k')(p \cdot k) + (p' \cdot k)(p \cdot k') - (p' \cdot k')(p \cdot k)]$$
$$= 256(p' \cdot k)(p \cdot k') \qquad (2.74)$$

Inserting Eq. (2.74) into Eq. (2.70), we can get the final averaged squared amplitude matrix as follows:

$$\frac{1}{12} \sum_{spin} \mathcal{M}^2 = \frac{1}{12} \cdot \frac{g_W^4}{64(q^2 - M_W^2)^2} \cdot 256(p' \cdot k)(p \cdot k')$$
$$= \frac{1}{3} \left(\frac{g_W^2}{q^2 - M_W^2} \right)^2 (p' \cdot k)(p \cdot k') \qquad (2.75)$$

Figure. 2.7 shows the center of mass frame (the W boson rest frame). In this frame, the angle between the outgoing muon and the incident proton with the positive longitudinal momentum is θ^*. With the notations defined in the figure, we get $p \cdot k' = p' \cdot k = E^2(1+\cos\theta^*)$, $q = p + p' = 2E$. The averaged squared amplitude in the mass of centre frame then is written as

$$\frac{1}{12}\sum_{spin}\mathcal{M}^2 = \frac{1}{3}\left(\frac{g_W^2}{q^2 - M_W^2}\right)^2 E^4 (1+\cos\theta^*)^2 \qquad (2.76)$$

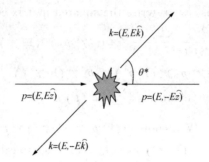

Figure 2.7 Center of mass frame for the process $q_i \bar{q}_j \to W^- \to \mu^- \bar{\nu}_\mu$

Finally, with the application of Eq. (2.64), the differential partonic W^- cross-section is written as follows:

$$\frac{d\sigma_{W^- \to \mu^- \bar{\nu}_\mu}^{partonic}}{d\Omega} = \frac{1}{2 E_{cm}^2} \frac{|\vec{k}|}{16\pi^2 E_{cm}} \frac{1}{12}\sum_{spin}\mathcal{M}^2$$

$$= \frac{1}{12} \frac{1}{64\pi^2} \frac{g_W^4 E^2}{(4E^2 - M_W^2)^2} (1+\cos\theta^*)^2 \qquad (2.77)$$

where $\vec{k} = E\hat{k}$, $E_{cm} = 2E$. This differential partonic W^- cross-section is maximal when the outgoing muon moves in the direction of the incoming proton with the positive longitudinal momentum.

With the integration on $d\Omega$, the total partonic W^- cross-section can be written as follows:

$$\sigma_{W^- \to \mu^- \bar{\nu}_\mu}^{partonic} = \frac{2 G_F^2 M_W^4 E^2}{9\pi (4E^2 - M_W^2)^2} \qquad (2.78)$$

where G_F is the Fermi constant which is defined as

$$\frac{G_F}{\sqrt{2}} = \frac{g_W^2}{8 M_W^2} \qquad (2.79)$$

The total partonic cross-section will blow up when the total energy, $2E$, reaches the W boson mass, M_W. This is due to the reason that we treat the W boson as a stable particle. In fact, it has a finite lifetime which will smear out the W boson mass. Thus the propagator is modified as

$$\frac{1}{q^2 - M_W^2} \to \frac{1}{q^2 - M_W^2 + i M_W \Gamma_W} \qquad (2.80)$$

where Γ_W is the decay rate (experimentally, $\Gamma_W = (2.085 \pm 0.042)$ GeV)[15]. If considering the quark mixing described in section 2.1.11, we should insert a factor $|V_{qq'}|^2$ into the total partonic cross-section. With these two adjustments, the total partonic cross-section in Eq. (2.78) becomes

$$\sigma_{W^- \to \mu^- \bar{\nu}_\mu}^{partonic} = \frac{2}{9\pi} \frac{G_F^2 M_W^4 E^2 |V_{qq'}|^2}{(4E^2 - M_W^2)^2 + M_W^2 \Gamma_W^2} \qquad (2.81)$$

Now let's turn to the calculation of the total hadronic W^- cross-section in pp collisions. As described in section 2.2.1, the hadronic cross-section is a convolution of the partonic cross-section and the PDF for the proton. The relation between the center of mass energy in the partonic process, $\hat{s}$, and the centre of mass energy in the hadronic process, s, is expressed as

$$\hat{s} = x_1 x_2 s \qquad (2.82)$$

With the application of Eq. (2.82), the total hadronic W^- cross-section is written as follows:

$$\sigma_{pp \to W^- \to \mu^- \bar{\nu}_\mu} = \frac{1}{18\pi} \int d x_1 d x_2 \sum_q$$
$$[f_{q_i/P_1}(x_1, Q^2) f_{\bar{q}_j/P_2}(x_2, Q^2) + f_{q_i/P_2}(x_1, Q^2) f_{\bar{q}_j/P_1}(x_2, Q^2)]$$
$$\frac{G_F^2 M_W^4 E^2 |V_{qq'}|^2 \cdot x_1 x_2 s}{(x_1 x_2 s - M_W^2)^2 + M_W^2 \Gamma_W^2} \qquad (2.83)$$

where $f_{q_i/P_1}(x_1, Q^2)$ and $f_{\bar{q}_j/P_2}(x_2, Q^2)$ are PDFs and q refers to the quark flavour. The second term which is proportional to $f_{q_i/P_2}(x_1, Q^2) f_{\bar{q}_j/P_1}(x_2, Q^2)$ comes from the interchange of q and $\bar{q}$ from the two incoming protons.

What we have discussed so far is the leading order (LO) production for the W boson in the pp collision at LHCb. There are also some higher order productions for the W bosons. They mainly arise from the gluon-quark interactions. Figure. 2.8 shows the Feynman diagrams which contribute at NLO.

2.2.3 Muon angular and p_T distributions

Before discussing the muon transverse momentum distribution, we first discuss the muon angular distribution in the W rest frame. Combing Eqs. (2.77) and (2.78), we get the negative muon angular distribution for the $W^- \to \mu^- \bar{\nu}_\mu$ process as follows:

$$\frac{1}{\sigma_{W^- \to \mu^- \bar{\nu}_\mu}^{partonic}} \frac{d \sigma_{W^- \to \mu^- \bar{\nu}_\mu}^{partonic}}{d \cos \theta^*} = \frac{3}{8} (1 + \cos \theta^*)^2 \qquad (2.84)$$

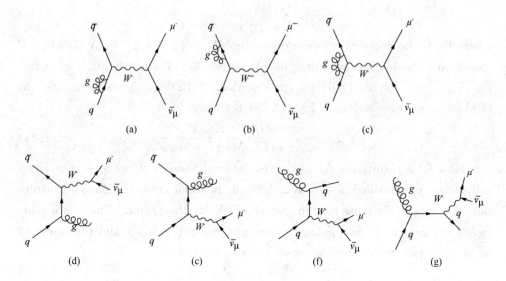

Figure 2.8 The NLO Feynman diagrams for the $q_i \bar{q}_j \to W$ process.
(a), (b) and (c) are the virtual gluon corrections.
(d) and (e) are real gluon corrections. (f) and (g) are quark-gluon scatterings

This equation is valid when the first proton, P_1, provides a down-type quark and the second proton, P_2, provides an up-type anti-quark. For the $W^+ \to \mu^+ \nu_\mu$ process, the positive muon angular distribution is

$$\frac{1}{\sigma_{W^+ \to \mu^+ \nu_\mu}^{\text{partonic}}} \frac{d\sigma_{W^+ \to \mu^+ \nu_\mu}^{\text{partonic}}}{d\cos\theta^*} = \frac{3}{8}(1-\cos\theta^*)^2 \qquad (2.85)$$

This equation is valid when P_1 provides an up-type quark and P_2 provides a down-type anti-quark. There are two other cases for the W production: one case is P_1 provides an up-type anti-quark and P_2 provides a down-type quark; the other case is P_1 provides a down-type anti-quark and P_2 provides an up-type quark. In the first case, the muon angular distribution is the same as Eq. (2.85). In the second case, the muon angular distribution is the same as Eq. (2.84).

These muon angular distributions could be explained schematically in terms of angular momentum conservation as follows. In Figure. 2.9(a), a left-handed d quark from P_1 annihilates with a right-handed $\bar{u}$ quark from P_2 and produces a W^- boson, as W bosons only couple to left-handed fermions or right-handed anti-fermions. The angular momentum is conserved in the collision, therefore the process with μ^- parallel to the direction of P_1 is allowed, but the process with μ^- parallel to the direction of P_2 is forbidden[28]. As a result, the angular distribution for the W^- boson is proportional to $(1-\cos\theta^*)^2$. Angular distributions in Figure. 2.9(b), 2.9(c) and 2.9(d) are explained in a similar way.

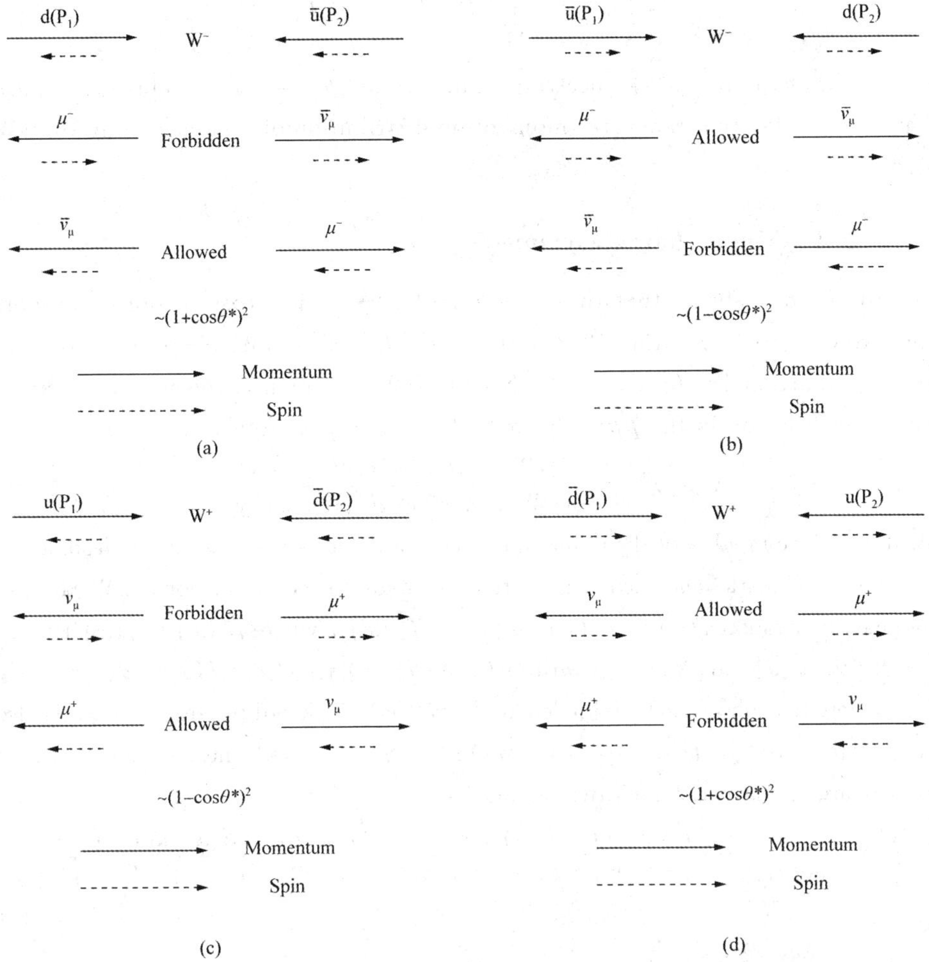

Figure 2.9 (a) and (b), preferred directions of muon leptons in W^- decay.

(c) and (d), preferred directions of muon leptons in W^+ decay.

Solid arrows show momentum directions, dashed arrows show spin directions

Combining Eqs. (2.84) and (2.85), we can get the averaged muon angular distribution for the W production as follows:

$$\frac{1}{\sigma_{W\to\mu\nu_\mu}^{\text{partonic}}} \frac{d\sigma_{W\to\mu\nu_\mu}^{\text{partonic}}}{d\cos\theta^*} = \frac{3}{8}(1+\cos\theta^*)^2 \tag{2.86}$$

In the center of the W boson mass frame, $\cos\theta^*$ can be written as follows:

$$\cos^2\theta^* = 1 - \frac{4 p_T^{*2}}{M_W^2} \tag{2.87}$$

where p_T^* is the muon transverse momentum in the W rest frame. Combining Eqs. (2.86) and (2.87), we can get the muon p_T^* distribution as follows[7]:

$$\frac{1}{\sigma_{W\to\mu\nu_\mu}^{\text{partonic}}} \frac{d\sigma_{W\to\mu\nu_\mu}^{\text{partonic}}}{dp_T^{*2}} = \frac{3}{2 M_W^2} \left(1 - \frac{4 p_T^{*2}}{M_W^2}\right)^{-\frac{1}{2}} \left(1 - \frac{2 p_T^{*2}}{M_W^2}\right) \quad (2.88)$$

It shows that the muon p_T^* spectrum peaks around $p_T^* = M_W/2$. This is a typical characteristic for the transverse momentum distribution of muons coming from W bosons.

2.2.4 Muon charge asymmetry

Eq. (2.84) shows that the differential $W^- \to \mu^- \bar{\nu}_\mu$ cross-section is proportional to $(1+\cos\theta)^2$. Eq. (2.85) shows the differential $W^+ \to \mu^+ \nu_\mu$ cross-section is proportional to $(1-\cos\theta)^2$. Now let's have a look at the charge asymmetry of the W boson in the pp collision[29]. It is defined as follows:

$$A_{+-}^W(y^W) = \frac{d\sigma(W^+)/dy^W - d\sigma(W^-)/dy^W}{d\sigma(W^+)/dy^W + d\sigma(W^-)/dy^W} \quad (2.89)$$

where y^W is the rapidity of the W boson in a lab frame. It is assumed that at leading order only u, d quarks and their anti-quarks contribute to the productions of W bosons. We take the notation $u(x) = f_u(x)$ $(d(x) = f_d(x))$ where $f_u(x)$ $(f_d(x))$ is the PDF for the $u(d)$ quark and we write $u(x) = u_V(x) + u_S(x)$ $(d(x) = d_V(x) + d_S(x))$. Here the subscripts V and S denote the valance quark and the sea quark. It is also assumed that $\bar{u}_S = u_S(x) = \bar{d}_S(x) = d_S(x) = S(x)$. With these notations, the W boson charge asymmetry is written as follows:

$$A_{+-}^W(y^W) = \frac{u_V(x_1) S(x_2) + S(x_1) u_V(x_2) - d_V(x_1) S(x_2) - S(x_1) d_V(x_2)}{u_V(x_1) S(x_2) + S(x_1) u_V(x_2) + d_V(x_1) S(x_2) + S(x_1) d_V(x_2) + 4 S(x_1) S(x_2)}$$

$$(2.90)$$

where $x_1 = \frac{M_W}{\sqrt{s}} \exp(y^W)$, $x_2 = \frac{M_W}{\sqrt{s}} \exp(-y^W)$.

Unfortunately it is very difficult to measure the W boson charge asymmetry, as the neutrino is undetected in the final state. What can be measured is the muon charge asymmetry, $A_{+-}^\mu(y^\mu)$. It is defined as follows:

$$A_{+-}^\mu(y^\mu) = \frac{d\sigma(\mu^+)/dy^\mu - d\sigma(\mu^-)/dy^\mu}{d\sigma(\mu^+)/dy^\mu + d\sigma(\mu^-)/dy^\mu} \quad (2.91)$$

where y^μ is the muon rapidity in the lab frame. The relation between y^μ and the muon rapidity in the W boson rest frame, y^*, is as follows:

$$y^\mu = y^W + y^* = y^W + \frac{1}{2}\ln\frac{1+\cos\theta^*}{1-\cos\theta^*} \quad (2.92)$$

According to Eq. (2.87), there are two solutions for $\cos\theta^*$ in $0 \leqslant p_T^* \leqslant M_W/2$: a positive one and a negative one, which correspond to positive and negative y^*. The fractions of momenta carried by the quarks, x_1 and x_2, then are written as follows:

$$x_1^{\pm} = x_0 \exp(+y_\mu) k^{\pm 1}, \quad x_2^{\pm} = x_0 \exp(-y_\mu) k^{\mp 1} \quad (2.93)$$

where $x_0 = (M_W/\sqrt{s})$, $k = \left(\dfrac{1+|\cos\theta^*|}{1-|\cos\theta^*|}\right)^{\frac{1}{2}} > 1$, $x_1^+ > x_1^-$ and $x_2^+ < x_2^-$ (see Figure. 2.10(a)). The muon charge asymmetry in Eq. (2.91) is written in terms of the quark and anti-quark PDFs as follows:

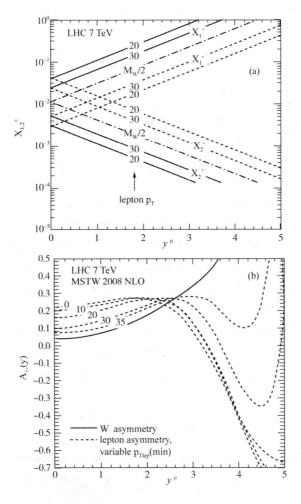

Figure 2.10 (a) $x_1^{\pm}$ and $x_2^{\pm}$ as a function of the muon rapidity. (b) Muon charge asymmetries with different p_T minimum cuts. Taken from Ref.[29]

$$A^\mu_{+-}(y^\mu) = \frac{A\ (1-\cos\theta^*)^2 + B\ (1+\cos\theta^*)^2}{C\ (1-\cos\theta^*)^2 + D\ (1+\cos\theta^*)^2} \tag{2.94}$$

where

$$\begin{cases} A = u_V(x_1^+)S(x_2^+) + u_V(x_1^-)S(x_2^-) - S(x_1^+)d_V(x_2^+) - S(x_1^-)d_V(x_2^-) \\ B = S(x_1^+)u_V(x_2^+) + S(x_1^-)u_V(x_2^-) - d_V(x_1^+)S(x_2^+) - d_V(x_1^-)S(x_2^-) \\ C = u_V(x_1^+)S(x_2^+) + u_V(x_1^-)S(x_2^-) + S(x_1^+)d_V(x_2^+) + S(x_1^-)d_V(x_2^-) \\ \quad + 2(S(x_1^+)S(x_2^+) + S(x_1^-)S(x_2^-)) \\ D = S(x_1^+)u_V(x_2^+) + S(x_1^-)u_V(x_2^-) + d_V(x_1^+)S(x_2^+) + d_V(x_1^-)S(x_2^-) \\ \quad + 2(S(x_1^+)S(x_2^+) + S(x_1^-)S(x_2^-)) \end{cases}$$

$$\tag{2.95}$$

Let's first have a look at the case with $y^\mu = 0$. In this case, $x_1^+ = x_2^- = x_0 k \equiv X$, $x_1^- = x_2^+ = x_0/k \equiv x$ and $X/x = k^2 \geqslant 1$ (see Figure. 2.10(a)). $X \gg x$ when the muon p_T is small. If X is not too close to 1, we can ignore the term of $V(x)S(X)$ and only consider the term $V(X)S(x)$. Then the muon charge asymmetry will approximately become

$$A^\mu_{+-}(y^\mu = 0) \approx \frac{u_V(X) - d_V(X)}{u_V(X) + d_V(X) + 2S(X)} \tag{2.96}$$

It reflects the difference between the u and d valence quarks. Figure. 2.10 (b) shows the muon charge asymmetry with different muon minimum p_T cuts. We find that the muon charge asymmetry with $y^\mu = 0$ goes up when the muon minimum p_T decreases. This could be approximately explained as follows. Figure. 2.5 shows that $u_V(X) - d_V(X)$ increases when X goes up at small X. When X goes up, k will go up, $\cos\theta$ will also go up and the muon p_T will go down.

Now let's have a look at the case with $y^\mu > 0$. When y^μ goes up from 0, x_1^+ will go to 1 and $x_{1,2}^-$ becomes the dominant contribution. We only take the term $V(x_1^-)S(x_2^-)$ into account as $x_1^- \gg x_2^-$ in this region. So the muon charge asymmetry becomes

$$A^\mu_{+-}(y^\mu > 0) \approx \frac{u_V(x_1^-)(1-\cos\theta^*)^2 - d_V(x_1^-)(1+\cos\theta^*)^2}{u_V(x_1^-)(1-\cos\theta^*)^2 + d_V(x_1^-)(1+\cos\theta^*)^2} \tag{2.97}$$

When $p_T^* \to M_W/2$, $\cos\theta^* \to 0$ and the muon charge asymmetry becomes

$$A^\mu_{+-}(y^\mu > 0)_{p_T^* = M_W/2} \approx \frac{u_V(x_1^-) - d_V(x_1^-)}{u_V(x_1^-) + d_V(x_1^-)} \tag{2.98}$$

It is always positive as we can see in the Figure. 2.5. $u_V > d_V$ is valid for all x. When the muon's p_T^* is small or moderate, $(1+\cos\theta^*)^2 \gg (1-\cos\theta^*)^2$ and the muon charge asymmetry becomes negative.

2.2.5 Muon charge ratio

The muon charge ratio is defined as follows

$$R^{\mu}_{+-}(y^{\mu}) = \frac{d\sigma(\mu^+)/dy^{\mu}}{d\sigma(\mu^-)/dy^{\mu}} \quad (2.99)$$

We can rewrite it in terms of the PDFs as follows

$$R^{\mu}_{+-}(y^{\mu}) = \frac{A'(1-\cos\theta^*)^2 + B'(1+\cos\theta^*)^2}{C'(1-\cos\theta^*)^2 + D'(1+\cos\theta^*)^2} \quad (2.100)$$

where

$$\begin{cases} A' = u_V(x_1^+)S(x_2^+) + u_V(x_1^-)S(x_2^-) + S(x_1^+)S(x_2^+) + S(x_1^-)S(x_2^-) \\ B' = S(x_1^+)u_V(x_2^+) + S(x_1^-)u_V(x_2^-) + S(x_1^+)S(x_2^+) + S(x_1^-)S(x_2^-) \\ C' = S(x_1^+)d_V(x_2^+) + S(x_1^-)d_V(x_2^-) + S(x_1^+)S(x_2^+) + S(x_1^-)S(x_2^-) \\ D' = d_V(x_1^+)S(x_2^+) + d_V(x_1^-)S(x_2^-) + S(x_1^+)S(x_2^+) + S(x_1^-)S(x_2^-) \end{cases} \quad (2.101)$$

Let's first have a look at the case with $y^{\mu} = 0$. If X is not too close to 1, we can ignore the term of $V(x)S(X)$ and only consider the term $V(X)S(x)$. Then the muon charge ratio will approximately become

$$R^{\mu}_{+-}(y^{\mu}=0) \approx \frac{u_V(X) + 2S(X)}{d_V(X) + 2S(X)} \quad (2.102)$$

It reflects the ratio between the u and d valence quarks.

Now let's have a look at the case with $y^{\mu} > 0$. When y^{μ} goes up from 0, x_1^+ will go to 1 and $x_{1,2}^+$ becomes the dominant contribution. We only take the term $V(x_1^-)S(x_2^-)$ into account as $x_1^- \gg x_2^-$ in this region. So the muon charge ratio becomes

$$R^{\mu}_{+-}(y^{\mu} > 0) \approx \frac{u_V(x_1^-)(1-\cos\theta^*)^2}{d_V(x_1^-)(1+\cos\theta^*)^2} \quad (2.103)$$

When $p_T^* \to M_W/2$, $\cos\theta^* \to 0$ and the muon charge ratio becomes

$$R^{\mu}_{+-}(y^{\mu} > 0)_{p_T^* = M_W/2} \approx \frac{u_V(x_1^-)}{d_V(x_1^-)} \quad (2.104)$$

It is always greater than 1 since $u_V > d_V$ is valid for all x (see Figure. 2.5). When the muon's p_T^* is small or moderate, $(1+\cos\theta^*)^2 \gg (1-\cos\theta^*)^2$ and the muon charge ratio becomes smaller than 1.

2.2.6 Motivations for measuring $\sigma_{W \to \mu\nu_{\mu}}$ at LHCb

LHCb is a single arm forward spectrometer specially designed to study b-physics. It covers the pseudo-rapidity, η, in the range $1.9 < \eta < 4.9$. Of this, $1.9 < \eta < 2.5$ is common to ATLAS and CMS while $2.5 < \eta < 4.9$ is unique to LHCb. Thus, LHCb provides complementary measurements of electroweak physics to those performed by ATLAS and CMS. Figure. 2.11(a) shows the probed regions in the (x, Q^2) space for different experiments at $\sqrt{s} = 7$ TeV. Two distinct probed regions for LHCb are indicated as the grey areas. In a high x region, PDFs are well constrained by previous experiments. In a low x region, the W cross-section measurement in LHCb probes PDFs in a region down to $x \approx 10^{-4}$ at a squared energy scale around $Q^2 = 10^4 \text{GeV}^2$. As discussed in section 2.2.1, the hadronic cross-section can be decomposed into two parts: one part is the partonic cross-section, which is predicted by the standard model; the other part is the PDFs.

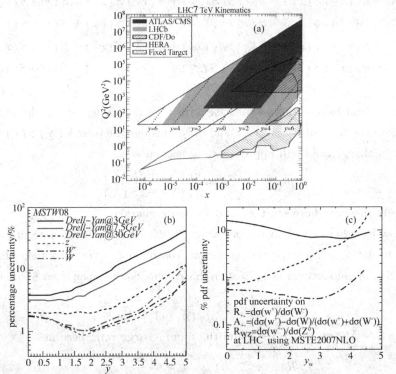

Figure 2.11 (a), probed regions in the (x, Q^2) space for different experiments at $\sqrt{s} = 7$ TeV. The grey areas show the probed regions in LHCb. Taken from Ref.[30]. (b), percentage uncertainties due to PDF errors on the cross-section as a function of the boson rapidity. (c), percentage uncertainties due to PDF errors on the cross-section ratio and charge asymmetry as a function of the boson rapidity. Taken from Ref.[31]

Figure. 2.11(b) shows percentage uncertainties due to PDF errors on the cross-section as a function of the boson rapidity with the MSTW08 PDF set. These uncertainties are calculated to NLO with the MCFM generator[32]. The black dash-dot curve is for the W^+ boson while the grey dash-dot one is for the

W boson. From the plot we can see that uncertainties due to PDF errors dominate at large rapidities. In kinematic regions where uncertainties due to PDF errors are low, a precise measurement of W boson cross-sections can test the standard model to a precision of 1%. In kinematic regions where uncertainties due to PDF errors are high, the measurement can reduce uncertainties on PDFs.

Figure. 2.11(c) shows percentage uncertainties due to PDF errors on the cross-section ratio and charge asymmetry as a function of the W boson rapidity. The dash curve shows the uncertainty for the W cross-section ratio. The solid curve shows the uncertainty on the W cross-section charge asymmetry. From the plot we can see that the uncertainty on the W cross-section ratio, R_{+-}, grows strongly at high rapidities. The measurement of the R_{+-} at LHCb can constrain the ratio between u and d valence quarks at a large x. The W cross-section charge asymmetry is predicted with the least precision. Its measurement at LHCb can constrain the difference between u and d valence quarks[33].

2.3 Parton shower and hadronization

As discussed in section 2.1.6, QCD predicts asymptotic freedom. Figure. 2.12 shows the evolution of the coupling constant, α_s, with the energy scale Q. In high-energy regions, α_s is small and the partonic cross-section can be calculated in a perturbative way. When the squared energy scale, Q^2, goes down to a threshold known as Λ_{QCD} ❶, α_s becomes large and the cross-section predictions are divergent. We can not use the perturbative method to do the cross-section calculation any more. Fortunately there are two approximate methods that can be applied to deal with this case. They are the parton shower and hadronization.

2.3.1 Parton shower

In section 2.2.2, we described the $W \to \mu \nu_\mu$ cross-section calculation at LO. As shown in the end of that section, the cross-section can be calculated to a higher order. Complete perturbative calculations for this cross-section have been performed to the next-to-next-leading-order (NNLO) in recent years[35]. However there are still some missing higher order terms, which will contribute to the overall cross-section. The emission of gluons or photons from initial and final state

---

❶ $\Lambda_{QCD} = 300$ MeV^2.

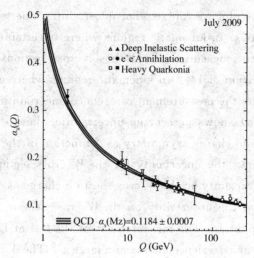

Figure 2.12 α_s as a function of the energy scale Q. Taken from Ref.[34]

particles play an important role in the higher order corrections to the overall cross-section. These emissions can be taken into account using the parton shower model. This model allows the parton to branch into a lower energy one with the radiation of a quark, gluon or photon. The radiation can occur before and after the hard process. They are known as initial state radiation (ISR) and final state radiation (FSR). In both ISR and FSR, the structure is given by a branching, $a \rightarrow bc$, such as $q \rightarrow q\gamma, q \rightarrow qg, g \rightarrow gg$ and $g \rightarrow q\bar{q}$. Each of these branchings is characterized by a splitting kernel, $P_{a \rightarrow bc}(z)$ where z is the fraction of momentum carried by the daughter, b. The branching rate is proportional to the integral, $\int P_{a \rightarrow bc}(z)dz$. Once the shower is formed, the daughters b and c will in turn branch, and so on. In the PYTHIA Monte Carlo generator (see section 2.4.1), the shower evolution stops when the energy of quarks and gluons reaches a cut-off energy, typically around 1 GeV for the QCD branching.

2.3.2 Hadronization

After the parton shower, there are free quarks and gluons which carry color. There are also some colored partons from the dissociated proton remnants. Due to the postulated color confinement, these colored quarks, gluons or partons can not exist individually. They must be converted into observable hadronic particles. This process is called hadronization. There are two ways for hadronization to be implemented in a Monte Carlo generator: the string fragmentation model[38] and cluster model[39]. They are described as follows.

String fragmentation model

The string fragmentation model is a phenomenological model for hadronization. It treats gluons as field lines. These gluons are attracted to each other due to their self-interactions and they form a string of the strong color field between two colored quarks, q and $\bar{q}$. The energy stored in the strong colour field increases linearly with the separation between these two quarks. When the stored energy exceeds the on-shell mass of the $q\bar{q}$ pair, the string may break by a production of a new $q'\bar{q}'$ pair from the vacuum. After the splitting, two color-singlet systems, $q\bar{q}'$ and $q'\bar{q}$, are formed. If the stored energies in these two new string pieces are large enough, further splittings will occur. The splitting process is assumed to proceed until only on-mass-shell colorless hadrons remain.

Cluster model

The cluster model describes that colored gluons are split into $q\bar{q}$ pairs. Color singlet clusters are formed by combining neighboring quarks and anti-quarks. Finally the clusters are fragmented into hadrons based on the cluster mass. As we know the whole mass spectrum of hadrons, we can compare the mass of the cluster to that of hadrons. If the cluster's mass is low, it may decay into a light single hadron whose mass is close to the cluster's mass. If the cluster is massive, then it may decay into two hadrons.

2.4 Monte Carlo event generators

Monte Carlo event generators are important to high energy particle physics. They are software applications which simulate event features of signal processes and their backgrounds. A number of different generators exist. They provide predictions for a variety of physical observables and these observables are compared to experimentally measured variables. In this section, we will describe three Monte Carlo event generators: PYTHIA, POWHEG, and MCFM.

2.4.1 PYTHIA

PYTHIA[36] is a general Monte Carlo generator which simulates processes at LO. Users can apply different PDF sets and phase-space cuts to these simulated processes. The string fragmentation model is used in PYTHIA to perform hadronization. An event table is produced after the simulation. It contains the particle type, momentum and charge information for each particle produced in each event. LHCb simulation package takes PYTHIA 6.4 as its default generator.

2.4.2 POWHEG

POWHEG[37] is a Monte Carlo generator which simulates hard scattering processes at NLO. As it is mainly dedicated to the simulation of vector boson production, LHCb does not utilize it as a default generator. In POWHEG, users can apply different PDF sets and phase-space cuts to generate events. In order to take into account the effects of ISR and FSR, POWHEG feeds these events to some modern shower Monte Carlo programs such as PYTHIA.

2.4.3 MCFM

MCFM[32] is a parton-level Monte Carlo generator. It is utilized to calculate cross-sections for various femtobarn-level processes at hadron-hadron colliders. For most processes, it evaluates matrix elements at NLO and incorporates full spin correlations. In MCFM, different PDF sets and phase-space cuts can be applied.

2.5 $\sigma_{W \to l\nu}$ and A^l_{+-} measurements at GPDs

W bosons can decay into charged leptons and neutrinos. The neutrinos are not detected experimentally in hadron colliders, only the charged leptons or their decay products are observed in the final states. As the muon or electron in the $W \to l\nu\,(l=e,\mu)$ decay process tends to be isolated and this muon or electron is easily reconstructed, the $W \to l\nu$ cross-section and its charge asymmetry measurements are also performed in several general particle detectors (GPDs), including the CDF detector at Tevatron, the ATLAS and CMS detectors at the LHC.

A $W \to \mu\nu$ cross-section measurement was performed with 72.0 pb^{-1} of data collected by the CDF collaboration in proton-antiproton ($p\bar{p}$) collisions at $\sqrt{s} =$ 1.96 TeV in 2005[40]. The pseudo-rapidity range of the CDF detector is $|\eta|<3.0$. In order to select the $W \to \mu\nu$ events, the muon's p_T should be greater than 20 GeV/c and the missing transverse energy, E_T ❶, for the neutrino should be greater than 20 GeV. The $W \to \mu\nu$ cross-section in the fiducial phase-space defined by these requirements is found to be $2768 \pm 16 \pm 64 \pm 166$ pb, where the first uncer-

---

❶ E_T is the transverse energy of a calorimeter tower and it is defined as $E_T = E\sin\theta$, where θ is the polar angle measured respect to the proton beam axis.

tainty is statistical, the second is systematic and the third is due the luminosity determination. The precision of this W cross-section measurement is 6.5%.

In 2009, Tevatron performed a direct measurement of the W production charge asymmetry as a function of the W boson rapidity, y_W, with 1 fb^{-1} of data collected by the CDF detector in $p\bar{p}$ collisions at $\sqrt{s} = 1.96\text{TeV}$[41]. Before this direct measurement, the W charge asymmetry was also measured as a function of the lepton pseudo-rapidity from the $W \to l\nu$ decay by the CDF and D0 collaborations[42,43]. However, as the lepton charge asymmetry is a product of the direct W boson charge asymmetry and the vector and axial vector (V-A) asymmetry from the W decays, and these two asymmetries will cancel at large η, the constrain on the proton PDF with the lepton charge asymmetry measurement will be weaker and complicated. This direct W production charge asymmetry measurement was performed with the $W \to e\nu$ decay. The $W \to e\nu$ events are identified by a large missing transverse energy, E_T, for neutrinos and a large transverse energy for isolated electrons. In order to determine y_W, the neutrino's longitudinal momentum should be known. This is done by constraining the $e\nu$ mass to that of the W boson. This W production charge asymmetry measurement is consistent with NLO and NNLO theoretical predictions. The precision of this measurement is about 15% at low y_W bins and decreases to about 4% at high y_W bins.

ATLAS made the W cross-section and its charge asymmetry measurements in the electron and muon channels with 35 pb^{-1} of data collected in 2010 in pp collisions at $\sqrt{s} = 7$ TeV[44]. The results with these two channels are combined together. The ATALS detector covers a range of $|\eta| < 2.5$. The measurement in the muon channel covers the whole η range, while in the electron channel, the range $1.37 < |\eta| < 1.52$ is excluded. The requirements applied in this measurement are that the lepton's p_T is greater than 20 GeV/c, the missing energy for the neutrino is greater than 25 GeV and the transverse mass for the lepton-neutrino system, m_T, is greater than 40 GeV/c^2. The combined W^+ (W^-) cross-section in the fiducial phase-space defined by these cuts is $3.110 \pm 0.008 \pm 0.036 \pm 0.106 \pm 0.004$ ($2.017 \pm 0.007 \pm 0.028 \pm 0.069 \pm 0.002$) nb, where the first uncertainty is statistical, the second is systematic, the third is due to the luminosity determination, the last is the uncertainty due to extrapolating the fiducial phase-space in the electron channel to that in the muon channel. The precision of the combined total W^+ (W^-) cross-section measurement is about 3.6% (3.7%). The precision of the combined differential W cross-section charge asymmetry as a function of the lepton's pseudo-rapidity ranges between 4% and 8%.

The W cross-section is also measured by the CMS collaboration in the electron and muon channels[45]. They utilized 36 pb^{-1} of data collected in 2010 in pp collisions at $\sqrt{s}=$ 7 TeV. The results with these two channels are also combined together. The pseudo-rapidity range covered by the CMS detector is $|\eta|<2.5$. The requirements applied on the electron channel are that the energy of the electron should be greater than 25 GeV and the electron's pseudo-rapidity should be in the range $|\eta|<1.44$ or $1.57<|\eta|<2.5$. The requirements applied on the muon channel is that the muon's p_T should be greater than 25 GeV/c and its pseudo-rapidity is in the range $|\eta|<2.1$. This acceptance factor is defined as the number of W events in the fiducial phase-space divided by that of W events in the whole phase-space, and is evaluated by POWHEG. The combined W^+ (W^-) cross-section in the whole phase-space is $6.04\pm0.02\pm0.06\pm0.08\pm0.24$ ($4.26\pm0.01\pm0.04\pm0.07\pm0.17$) nb, where the first uncertainty is statistical, the second is systematic, the third is the theoretical uncertainty which affects the acceptance factor determination, the last is due to the luminosity determination. The precision of the W^+ (W^-) cross-section measurement is 4.3% (4.4%). The W cross-section charge asymmetry as a function of the lepton pseudo-rapidity is measured in the muon and electron channels separately with the lepton's p_T greater than 25 and 30 GeV/c[46]. The precision of the differential W cross-section charge asymmetry measurement is less than 1.6% for both channels.

In this book, the W cross-section and its charge asymmetry measurements are performed in the LHCb detector with 37 pb^{-1} of data collected in 2010 in the pp collisions at $\sqrt{s}=7$ TeV. As described in section 2.2.6, the LHCb detector covers a range of $1.9<|\eta|<4.9$. The first range $1.9<|\eta|<2.5$ is common to those covered by the ATLAS and CMS detectors while the second $2.5<|\eta|<4.9$ is unique in LHCb. The first range is an overlap region between LHCb and ATLAS (CMS), thus it provides a comparison between the measurements in these three detectors. With the second range, LHCb provides complementary measurements to those performed at ATLAS and CMS. In LHCb the W cross-section and its charge asymmetry are only measured in the muon channel. It requires that the muon's p_T should be greater than 20 GeV/c and its pseudo-rapidity should be in the range $2.0<|\eta|<4.5$. As there is saturation in the energy readout of the cell in the electromagnetic calorimeter❶, and this saturation pre-

---

❶ The E_T threshold of this saturation is around 10 GeV [78].

vents the measurements of the missing energy for the neutrino as well as the transverse mass of the muon-neutrino system. As a result, we could not directly compare the measurement results in LHCb to the results in ATLAS or CMS. However, this comparison is possible if we apply a correction factor for extrapolation to the LHCb result. For the W cross-section, the correction factor for the LHCb result extrapolation is defined as the predicted cross-section for the fiducial phase-space of ATLAS or CMS divided by the predicted cross-section in the fiducial phase-space of LHCb[47]. For the W cross-section charge asymmetry, the correction factor for the LHCb result extrapolation is defined as the difference between the asymmetries predicted in the LHCb and ATLAS (CMS) fiducial phase-spaces. Section 5.15 shows the W cross-section and its charge asymmetry after extrapolation in LHCb agree well with those at ATLAS❶.

---

❶ In section 5.15, we do not compare the W cross-section and its charge asymmetry measurements in LHCb with the results in CMS, as in Ref. [47] the correction factors for extrapolations from $p_T^l=$ 20 GeV/c to 25 GeV/c and 30 GeV/c are not available.

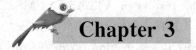

Chapter 3

Experimental environment

The Large Hadron Collider (LHC) is the largest and highest energy particle accelerator in the world. It was built by the European Organization for Nuclear Research (CERN) from 2001 to 2008. It allows physicists to test the standard model in a high-energy region. It answers the question of whether the masses of elementary particles are generated by the Higgs mechanism via electroweak symmetry breaking or not. Some new physics such as super-symmetry[48], an extension to the standard model, and extra dimensions are also searched for at the LHC. LHCb[49] is one of four main experiment detectors at the LHC. It is dedicated to B-physics. It measures the parameters of CP violation in B-hadron interactions. Rare decays of B-hadrons such as $B_s^0 \to \mu^+ \mu^-$ are also searched for at LHCb[50]. The branching fraction of this rare decay is sensitive to new processes or new heavy particles, thus measuring this branching fraction provides a method to look for new physics. Besides B-Physics, electroweak physics in the forward region is also studied in LHCb[51, 52, 53].

3.1 LHC

The LHC is built in a circular tunnel, which is situated 100 m below the ground. The tunnel width is 3.8 m and its circumference is 27 km. This tunnel was originally built for the Large Electron-Positron (LEP) collider in CERN. Inside the tunnel there are two adjacent parallel beam pipes, which intersect at four points. Four main detectors ATLAS, LHCb, CMS and ALICE are built around these four points. Each beam pipe contains a proton beam. The two beams travel in opposite directions to each other along the tunnel. In order to keep the particle beams on the circular path, about 1232 dipole magnets are installed around the beam pipes. Another 392 quadrupole magnets are installed at the intersection points. They are used to keep the beam focused. In total there are over 1600 magnets. These magnets are superconducting and they are surrounded by approximately 96 tonnes of liquid helium. The helium is maintained at a temperature below 1.9 K and a pressure below 130 kPa. This extremely low temperature allows magnetic field strength strong enough to bend the high-energy particle beam. The protons are bunched together when travelling around the ring. At a peak operation, there will be 2808 bunches of protons circulating in the ring. In each bunch there are 115 billion protons. These bunches are separated by a time interval of 25 ns[54]. In 2010, the LHC ran the collision with a minimum bunch separation of 50 ns, and a maximum of 368 bunches per beam circulated in the ring[55].

3.1.1 LHC acceleration systems

Figure. 3.1 shows the LHC acceleration system. Before being injected into the main accelerator, the protons are accelerated by a series of systems. The first system is the linear particle accelerator (LINAC 2). It accelerates the protons to 50 MeV. After the first acceleration, the protons are fed to the Proton Synchrotron Booster (PSB). From there the protons are boosted to 1.4 GeV and then they are passed to the Proton Synchrotron (PS) which accelerates the protons to 25 GeV. The energy of protons is further increased to 450 GeV by the Super Proton Synchrotron (SPS). After these series of accelerations the protons are injected into the main ring. It takes a period of 20 minutes for the proton bunches to be accelerated to their peak colliding energy 7 TeV. Finally the proton bunches circulate around the ring for 10 to 24 hours and the collisions occur at the four intersection points: ATLAS, LHCb, CMS and ALICE.

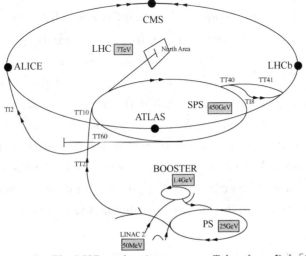

Figure 3.1 The LHC acceleration systems. Taken from Ref. [56]

3.1.2 LHC Luminosity

The instantaneous luminosity, L, is a measurement of the ability of a particle accelerator to produce a given number of interactions. It is a constant of proportionality between the number of events per second, dN_{EVT}/dt, and cross-section of a given process, σ_p:

$$\frac{dN_{EVT}}{dt} = L \cdot \sigma_p \qquad (3.1)$$

The cross-section, σ_p, can be interpreted as the probability of that given process to occur in a given condition. The designed luminosity for the LHC is around $10^{34}\,cm^{-2}s^{-1}$. A reduced luminosity around $10^{32}\,cm^{-2}s^{-1}$ is delivered to LHCb❶. In order to get the total number of events, N_{EVT}, for a given process, Eq. (3.1) is integrated over a period of time, T:

$$N_{EVT} = \sigma_p \cdot \int_0^T L\,dt \qquad (3.2)$$

where $\int_0^T L\,dt$ is called the integrated luminosity. It is measured in units of inverse barns, b^{-1}, where $1b^{-1} = 10^{24}\,cm^{-2}$. Some derived units of the luminosity can be found in this book, they are inverse picobarns, pb^{-1} ($1pb^{-1} = 10^{12}\,b^{-1}$), and inverse femtobarns, fb^{-1} ($1fb^{-1} = 10^{12}\,b^{-1}$).

Another expression for the luminosity is based on the beam parameters[57] and it is written as follows:

$$L = \frac{N_1 N_2 f N_b}{4\pi \sigma_x \sigma_y} \cdot R_\phi \qquad (3.3)$$

where N_1 and N_2 are the number of particles per bunch, N_b is the number of bunches per beam, f is the beam crossing frequency. σ_x and σ_y give the extensions of the bunches in the horizontal and vertical directions. R_ϕ is called the luminosity reduction factor which is due to the non-vanishing LHC beam crossing angle.

Based on the definition in Eq. (3.3), the luminosity can be determined by measuring the beam parameters. At LHCb, the beam parameters can be measured in two methods. One method is to use the Van der Meer Scan[58]. The other method is to use the Vertex Locator to measure the characteristics of beam-gas events near the interaction point[59]. LHCb gets an overall precision of 3.5% in the absolute luminosity determination when combining the two methods of measuring

---

❶ This luminosity reduction is achieved by tuning the proton beam-crossing angle.

the beam parameters.

3.1.3 LHC performance

The LHC started to accelerate protons for the first time on the 10th of September 2008. But 9 days later a serious accident involving a leakage of liquid Helium in the tunnel occurred. Due to this accident, beam collisions were delayed by 14 months. On the 30th of March 2010, the first collision with two 3.5 TeV beams took place after that accident. During the years of 2010 and 2011, the LHC operated at 3.5 TeV per beam smoothly. In 2012(2015) it operated at 4 (6.5) TeV per beam. It was shut down in 2013 and 2019 for upgrades. Figure. 3.2 shows the luminosity delivered to and recorded by LHCb in the years of 2010 and 2011 at 3.5 TeV per beam, and in the year of 2012 at 4 TeV per beam. In this book, 2010 collision data with an integrated luminosity of 37.1 pb^{-1} is used.

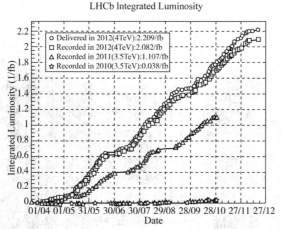

Figure 3.2 The integrated luminosity recorded by LHCb in 2010 (star points), 2011 (triangle points) and 2012 (square points). The delivered luminosity in 2012 is shown as the circle points.

Taken from Ref. [60]

3.2 LHCb

B-hadrons and the particles they decay into are predominantly produced close to the line of the beam pipe. This is reflected in the design of the LHCb detector. LHCb is a single arm spectrometer. The forward angular coverage of the LHCb detector is approximately from 10 mrad to 300 mrad in the vertical direction and from 10 mrad to 250 mrad in the horizontal direction. With this design, the LHCb detector covers the pseudo-rapidity, η, in the range $1.9 < \eta < 4.9$. Figure.

3.3 shows the layout of the LHCb detector. A right-handed coordinate system is adopted in this figure. Here the z axis is along the beamline direction, the y axis is along the vertical direction and the x axis is along the horizontal direction. The proton-proton collisions take place at the origin, (0,0,0). From the left to the right the instruments are the Vertex Locator, the dipole magnet, the RICH detectors, the tracking stations TT and T1 to T3❶, the Scintillating Pad Detector, the Pre-Shower, the Electromagnetic and Hadronic calorimeters, and the muon stations M1 to M5. In the following sections, each instrument in the LHCb detector will be described in detail.

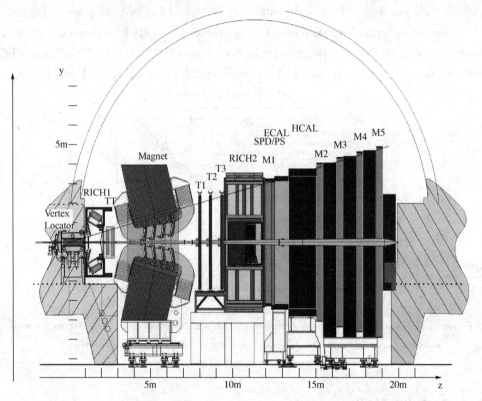

Figure 3.3 The LHCb detector layout. Taken from Ref.[49]

3.2.1 Vertex locator

The VEertex LOcator (VELO) is built around the pp interaction region. It provides measurements of particle trajectories near the interaction region. The

---

❶ The TT station is located upstream of the magnet, the three T stations, T1 to T3, are installed downstream of the magnet.

primary and secondary vertices are reconstructed and separated with these measurements. The VELO consists of 42 semi-circular modules positioned perpendicular to the beamline. These 42 modules lie in positions between $z = -18$ cm and $z = 88$ cm. Half of them are on one side of the beamline while another half of them are on the other side of the beamline. In each module there are two back-to-back silicon detectors: R and ϕ sensors. These sensors are utilized to measure the radial R and azimuthal ϕ coordinates of transversing particles. R and ϕ sensors are only 300 μm thick. The z coordinate of a particle is determined by the module position along the z axis. Figure. 3.4 shows the cross-section of the VELO silicon sensors in the (x, z) plane at $y = 0$. The interaction region is indicated by the grey area. It is located around $z = 0$ with a longitudinal uncertainty of 5.3 cm. R sensors are indicated by red solid lines while ϕ sensors are indicated by blue dashed lines. The bottom two plots in Figure. 3.4 show two states of the VELO: one is fully closed and occurs during data taking and the other one is fully open which happens during the beam injection. When the VELO is fully closed, there is an area of 8 mm in radius around the beamline for the passage of the beam. When the VELO is fully open, both VELO halves are retracted to a distance of 29 mm for safety reasons. In addition to VELO sensors, there are two pile-up veto stations. They are located upstream of the VELO. Each pile-up veto station consists of 2 modules. In each module, there is a R sensor only.

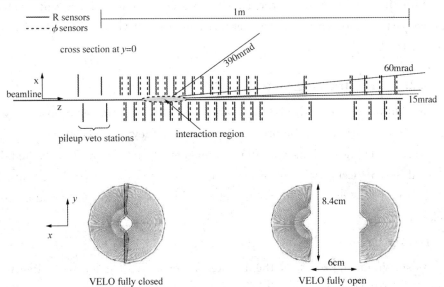

Figure 3.4 The VELO detector layout. Taken from Ref.[49]

3.2.2　The magnet system

In the LHCb experiment, a dipole magnet is used to bend charged particles. The momentum of a particle can be measured by examining the curvature of the particle trajectory. Figure. 3.5(a) shows the layout of the LHCb magnet system. This magnet system is located between the TT and tracking stations T1, T2 and T3. It consists of two saddle shaped coils mounted on a window-frame yoke. The opening angle of the magnet in the vertical direction is ± 250 mrad and in the horizontal direction it is ± 300 mrad. This design makes sure that the magnet system covers the entire LHCb acceptance. The main component of the magnetic field is along the y axis, thus particles moving in the z direction are bent in the xz plane. The direction of the magnetic field is changed periodically in order to control systematic effects of the detector during data taking. Figure. 3.5(b) shows the dependence of the magnetic field strength (B_y) on the position along the z axis for both polarities: magnet down and magnet up. This dependence ensures that the magnetic field strength is minimized in the VELO where a fast straight track finding is essential for the L0 trigger system. The integrated magnetic field along the z axis is

$$\int B \, dl = 4.2 \text{ Tm} \tag{3.4}$$

With the combination of the LHCb magnet system and tracking system, charged particle momenta can be measured up to 200 GeV/c at a precision of about 0.4%[61].

3.2.3　The tracking system

In addition to the VELO, there are five other tracking stations. Two of these tracking stations are before the magnet and they are labeled as TT. The other three tracking stations are after the magnet and labeled as T1, T2 and T3. Due to different occupancies in different regions, each of the tracking stations, T1, T2 and T3, is divided into two segments: an inner tracker (IT) and an outer tracker (OT).

Tracker turensis

As shown in Figure. 3.3, the Tracker Turensis(TT) is located between the magnet and RICH-1. It is placed inside a large light-tight, thermally and electrically insulated box, which is kept below a temperature of 5℃. The TT consists

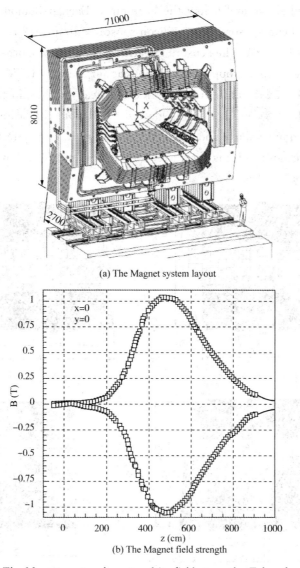

(a) The Magnet system layout

(b) The Magnet field strength

Figure 3.5 The Magnet system layout and its field strength. Taken from Ref.[49]

of two stations. In each station there are two detection layers. The detection layers in the first station are labeled as TTa while in the second station they are labeled as TTb. TTa are centred around $z = 232$ cm and TTb are centred around $z = 262$ cm. The readout strips in the first and fourth detection layers are vertical while in the second and third layers they are rotated by $\pm 5°$ with respect to the y axis. The stereo angle allows tracks to be reconstructed in three dimensions. Figure. 3.6 shows layouts of these four layers. Each layer is assembled with half modules located above and below the beampipe. Each half module consists of a row of seven silicon sensors. Each silicon sensor contains 512 readout strips with a strip pitch of 183 μm. The sensors in each half module are organized in a 4-3 or 4-2-1 configuration and grouped into three readout sectors, L, M and K. The half module with the 4-2-1 configuration is located near the beampipe while the half module with the 4-3 configuration is located away from the beampipe. There are 15 (17) half modules above and below the beampipe in each detector layer of TTa (TTb). The readout hybrids for all readout sectors are placed at the edge of TT sensors.

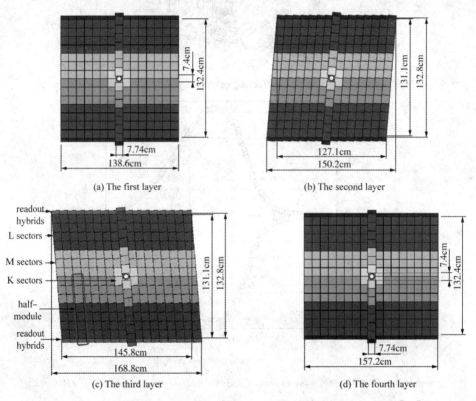

Figure 3.6 The layout of the four layers in the TT. Taken from Ref.[62]

The TT has been built for two reasons. Firstly, as the magnetic field in the TT station is not negligible, track momenta can be measured to a precision of 20% when hits in the TT are combined with the information in VELO. Although this measurement provides a less precise momentum than the measurement with other tracking stations, it is useful in the HLT1 trigger stage as the full tracking system (VELO-TT-T1-T2-T3) uses much more time than the tracking system with only VELO and TT. Secondly, the pseudo-rapidity coverage of the TT station is $2 \lesssim \eta \lesssim 4.5$, which is more restrictive than the rest of the tracking systems. In the offline analysis, due to low momenta, some tracks are bent outside the acceptance of other tracking systems, but these tracks are still in the acceptance of the TT station. As a result, the TT allows these tracks to be reconstructed. The TT station also allows trajectories of long-lived neutral particles (e. g. K_S^0) which decay outside the VELO to be reconstructed[63].

Inner tracker

The IT is a silicon strip detector in the center of the tracking stations T1, T2 and T3, thus it receives the highest flux of charged particles. In each of the IT stations there are four separated boxes arranged in a cross-shaped configuration around the beampipe. These boxes are light-tight, thermally and electrically insulated, and kept below a temperature of 5℃. Figure. 3.7(a) shows the overview of four IT detector boxes. In each box there are four detection layers. The first and last layers consist of vertical micro-strips. The second and third layers consist of micro-strips which are rotated by $\pm 5°$ with respect to the y axis. In each detection layer, there are two types of modules. The first type of module consists of a single silicon sensor and a readout hybrid and is located in the boxes above and below the beampipe. The other type of module consists of two silicon sensors and a readout hybrid and is located in the boxes on the left and right of the beampipe. In each silicon sensor there are 384 readout strips with a strip pitch of 198μm. Figure. 3.7(b) shows the layout of one layer with a vertical configuration in the second tracking station, T2.

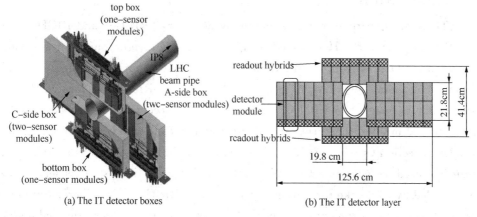

(a) The IT detector boxes   (b) The IT detector layer

Figure 3.7 The IT detector boxes and layer. The layer is in a vertical configuration and is taken from the second tracking station, T2. Taken from Ref.[49]

Outer tracker

The OT is a drift time detector. It is located in the remaining area of the tracking stations, T1, T2 and T3 (see Figure. 3.8(a)). In each OT there are four layers. The first and last layers consist of an array of vertical gas-tight straw-tube modules. The second and third layers consist of an array of straw-tube modules which are rotated by $\pm 5°$ with respect to the y axis. In each straw-tube module there are two staggered layers of 64 drift-tubes with inner diameters of 4.9 mm (see Figure. 3.8(b)). In each drift-tube, a 25 μm anode is surrounded by a gas-filled tube. The wall of the tube operates as a cathode. The anode consists of a gold-plated tungsten wire. The cathode consists of two foils: the inner foil is made of 40 μm carbon doped polymide; the outer foil is made of a 25 μm polymide layer with a 12.5 μm aluminium coating. The gas in the drift-tube is ionized when an energetic charged particle passes through the tube. The liberated electrons drift towards the anode of the drift-tube. An electron avalanche is initiated by the strong field strength around the anode wire. This avalanche is collected by the anode and results a detectable electric current as a signal. There is a delay between the time when the particle passes through the tube and the time when the signal is read out. This delay is known as the drift-time of the ionized electron and it can be used to determine the position where the charged particle traverses the detector. In order to get a fast drift-time ($<$50 ns) and a sufficient drift-coordinate resolution (200 μm), the gas in the tube is a mixture of Argon (70%) and CO_2 (30%)[49].

3.2.4 RICH

There are two ring-imaging Cherenkov detectors (RICH-1 and RICH-2) in the RICH system. RICH-1 is located upstream of the magnet and it has a wide acceptance which is from 25 mrad to 300 mrad in the horizontal direction and from 25 mrad to 250 mrad in the vertical direction. RICH-2 is located downstream of the magnet and it has a limited angular acceptance which is from 15 mrad to 120 mrad in the horizontal direction and from 15 mrad to 100 mrad in the vertical direction (see Figure. 3.3, 3.10(a), 3.10(b))[49]. The RICH system is in principle used to separate π from K in selected B-hadron decays. This separation is provided by Cherenkov radiation in RICH-1 and RICH-2. The Cherenkov radiation is emitted in a cone when a charged particle traverses a dielectric medium at a speed, v, greater than the speed of light in that medium, c/n, where n is the

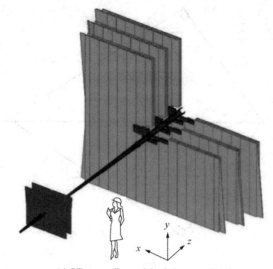

(a) OT straw-tlbe modules in layers and stations

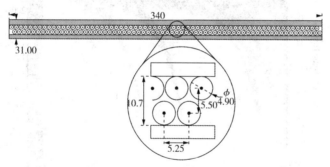

(b) Cross-section of a straw-tubes module in the OT detector

Figure 3.8 In the upper plot, the OT layers are in grey color. The IT and TT modules are shown in black color. Taken from Ref.[49]

refractive index of the medium and c is the speed of light in vacuum. Figure. 3.9 (a) shows the geometry of the Cherenkov radiation. The polar angle θ is the angle at which the photons are emitted and it is know as Cherenkov angle θ_c. This angle is determined as follows:

$$\cos \theta_c = \frac{1}{n\beta} \quad (3.5)$$

where $\beta = v/c$. The velocity of the charged particle, v, can be determined by measuring the polar angle, θ_c, of the emitted photon. Combining this velocity information with the momentum information from the tracking stations, we can determine the mass and type of the particle.

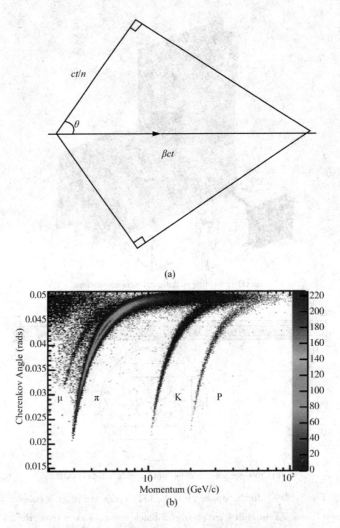

Figure 3.9 (a) The geometry of the Cherenkov radiation. (b) Cherenkov angle θ_C as a function of particle momentum for the C_4F_{10} radiator. Taken from Ref. [64]

In Eq. (3.5), $\cos\theta_C \leq 1$ requires that the velocity of the charged particle should exceed the threshold, $\beta_t = 1/n$, in order to initiate the Cherenkov radiation. In order to identify low momenta particles, a material with a large refractive index is needed. Conversely, in order to identify high momenta particles, a material with a small refractive index is needed. Thus, in order to identify particles over a large momentum range, two RICH detectors utilize three kinds of Cherenkov radiators: Aerogel, C_4F_{10} and CF_4. Figure. 3.9(b) shows the Cherenkov angle, θ_C, as a function of particle momentum with the C_4F_{10} radiator for isolated tracks selected in data. These tracks fall into distinct bands according to their masses. In RICH-1, there are two radiators, one is a 5 cm silica Aerogel which is located near the VELO exist window, the other one is the C_4F_{10} gas (see Figure. 3.10(a)). With these two radiators, RICH-1 can distinguish π and K over a momentum range of 1-60 GeV/c. In RICH-2, there is only one radiator, CF_4 gas. With this radiator, RICH-2 can distinguish π and K over a momentum range from 15 to 100 GeV/c. In both RICH detectors, the Cherenkov light is focused and reflected into an array of Hybrid Photon Detectors (HPDs) with a set of mirrors. When an incident photon hits a vacuum photo-cathode in HPDs, a photoelectron is emitted from the photo-cathode by the photoelectric effect and accelerated by an electric field onto a silicon anode. The photoelectron then is absorbed by the silicon anode, and this absorption results in the creation of electron-hole pairs. With the influence of the electric field, the electron and hole move in the opposite direction. This will generate electronic signals, which indicate a photon is detected.

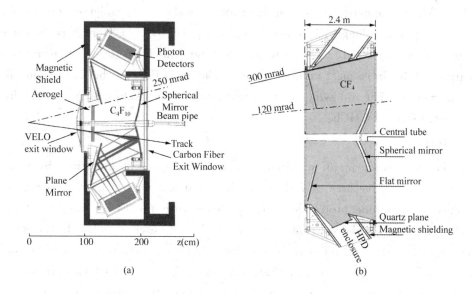

Figure 3.10 (a) Side view schematic layout of the RICH-1 detector. (b) Top view schematic layout of the RICH-2 detector. Taken from Ref.[49]

3.2.5 Calorimeters

The calorimeter system in LHCb consists of four calorimeters: a Pre-Shower (PRS), a Silicon Pad Detector (SPD), an electromagnetic calorimeter (ECAL) and a hadronic calorimeter (HCAL). It is located between the first and second muon stations (see Figure. 3.3). The calorimeter system identifies hadrons, electrons and photons and measures their energies and positions. The transverse energies of hadrons, electrons and photons can be used in the L0 trigger stage which makes a decision to accept or reject an event 4 μs after the interaction.

The calorimeter system is based on the detection of particle showering which happens when particles traverse the material. The showering can be separated into two types: one type is electromagnetic and the other type is hadronic. The electromagnetic showering happens in two processes: one process is Bremsstrahlung radiation in which a photon is emitted by a charged particle; the other process is pair production which is due to photon-nuclei interactions. The scale of electromagnetic showering is determined by the radiation length, X_0, over which a charged particle will lose 63% of its energy due to Bremsstrahlung. The energy loss due to Bremsstrahlung is inversely proportional to the mass squared of the charged particle. Hadronic showering happens when a traversing hadron interacts with nuclei of the material. The scale of hadronic showering is determined by the nuclear interaction length, λ_I, which is proportional to $A^{1/3}$. Here A is the mass number of the material. In order to get a shorter scale for hadronic showering, a material with a large mass number, iron, is chosen as the absorber in the hadronic calorimeter. The following sections will describe each calorimeter in the calorimeter system of LHCb.

SPD

The SPD is a plane of scintillator pads containing 12032 detection channels. It covers a sensitive area of 7.6 m in width and 6.2 m in height. In order to achieve a one-to-one projective correspondence with the ECAL segmentation, each SPD plane is subdivided into inner, middle and outer sections. In the inner section, there are 3072 cells, each of which is in an approximate 4×4 cm^2 cell dimension. In the middle section, there are 3584 cells, each of which is in an approximate 6×6 cm^2 cell dimension. In the outer section, there are 5376 cells, each of which is in an approximate 12×12 cm^2 cell dimension. Figure. 3.11(a) shows one quarter of the lateral segmentation of the SPD. The light generated in each cell is read out by a single wavelength-shifting (WLS) fibre that is coupled to a multianode photomultiplier tube (MAPMT) via clear plastic fibres. As no showering happens before the SPD, it only detects charged particles. This enables the SPD to distinguish between neutral and charged particles. Thus hits in the SPD can be utilized in the suppression of backgrounds from neutral pion decays $\pi^0 \to \gamma\gamma$ during the electron candidate selection.

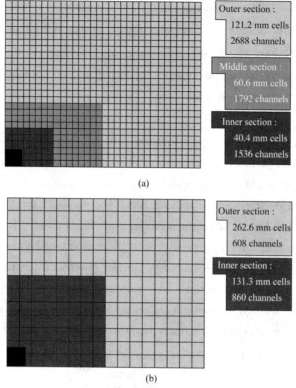

Figure 3.11 (a) One quarter of the lateral segmentation of the SPD/PS and ECAL. (b) One quarter of the lateral segmentation of the HCAL. Taken from Ref.[49]

PRS

The PRS is another scintillator plane, which is identical to the plane in the SPD. It is located downstream of the SPD. Between planes in the SPD and PRS, there is a lead converter with a 15 mm thickness, which corresponds to a radiation length of 2.5 X_0. The distance between centers of the PRS and SPD planes along the beam axis is 56 mm. The lead converter is thick enough to initiate an electromagnetic showering but insufficient to cause a significant hadronic showering. The electromagnetic showering detected by the PRS enables the differentiation of electrons and charged pions.

ECAL

The ECAL is a sampling calorimeter. It is located at a distance of 12.5 m from the interaction point along the z axis (see Figure. 3.3). The ECAL consists of 66 alternating layers of lead absorbers and scintillator tiles. The lead absorber is 2 mm thick, and the scintillator tile is 4 mm thick. The total thickness of the calorimeter corresponds to a radiation length of 25 X_0. As the hit density varies as a function of the distance from the beampipe, the ECAL is subdivided into three sections: the inner, middle and outer sections (see Figure. 3.11(a)). The experimental energy resolution for the ECAL is $\sigma_E/E = 8\%/\sqrt{E} \oplus 0.8\%$ [65], where E is the measured energy, the first term is the statistical uncertainty, the second term is the systematic uncertainty, and $\oplus$ denotes adding the two uncertainties in quadrature.

HCAL

The HCAL is also a sampling calorimeter. It is located at a distance of 13.33 m from the interaction point along the z axis (see Figure. 3.3). The dimension of the HCAL is 8.4 m in height, 6.8 m in width and 1.65 m in depth. It consists of alternating layers of iron absorbers and scintillator tiles. The iron is 16 mm thick and the scintillator tile is 4 mm thick. Unlike the scintillator tile in the SPD, PRS and ECAL, the HCAL scintillator tiles are placed parallel to the beampipe (see Figure. 3.12). The total thickness of the HCAL corresponds to a nuclear interaction length of 5.6 λ_I. The HCAL is subdivided into two sections: the inner and outer sections (see Figure. 3.11 (b)). In the inner section, the cell dimension is approximate 13×13 cm^2. In the outer section, the cell dimension is approximate 26×26 cm^2. The experimental energy resolution for the HCAL is $\sigma_E/E = (67)\%/\sqrt{E} \oplus (9)\%$ [65].

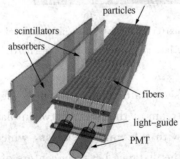

Figure 3.12 The internal cell structure of HCAL. The scintillator-absorber layer is oriented parallel to the beampipe. "PMT" refers to the photomultiplier tube. Taken from Ref.[49]

3.2.6 Muon system

The muon system is designed to provide fast information for the high-p_T muon trigger at the L0 trigger stage, the muon identification for the high-level trigger (HLT) and the offline analysis. It consits of five muon stations: M1, M2, M3, M4 and M5 (see Figure. 3.3). M1 is located in front of the calorimeter system, M2 to M5 are located downstream of the calorimeter system. The stations M2 to M5 are separated by 80 cm thick iron filters (see Figure. 3.13(a)). These iron filters absorb hadrons and electrons and thereby they reduce the possibility that these particles will be misidentified as muons. The total thickness of the muon system corresponds to 20 nuclear interaction lengths. In order to traverse the five muon stations in the muon system, a muon candidate should have a minimum momentum of 10 GeV/c. The first three muon stations have high spatial resolutions along the x axis in the bending plane and they are primarily built for momentum measurements of candidate muons with a resolution of 20%. The last two muon stations have limited spatial resolutions and they are primarily built for the identification of penetrating particles. The angular acceptance for the inner and outer of the muon system is 20 (16) mrad and 306 (258) mrad in the bending (non-bending) plane respectively. This acceptance covers about 20% of muons from inclusive b semi-leptonic decays.

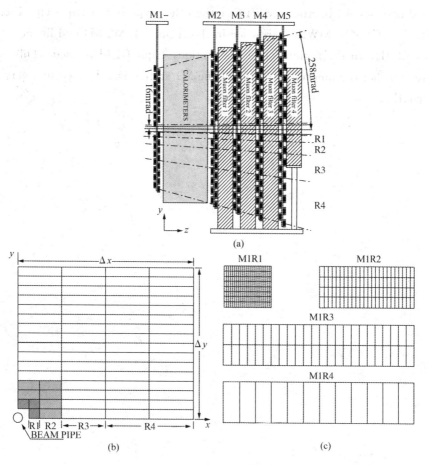

Figure 3.13 (a) Side view of the muon system. (b) Front view of a quadrant of a muon station. Each rectangle represents one chamber. (c) Division into logical pads of four chambers belonging to the four regions of the station M1. Taken from Ref.[49]

Each muon station is divided into four regions: R1 to R4. R1 is closet to the beampipe, R4 is the furthest away. Figure. 3.13(b) shows the layout of one muon station in the muon system. In each station there are 276 chambers. Each chamber is divided into pads. Figure. 3.13(c) shows the logical pad divisions of four chambers in the four regions of the station M1. The dimensions of the pads in the regions R1-R4 scale in the ratio of 1 : 2 : 4 : 8. The particle flux is highest in the region near the beampipe and is smaller in the region far away from the beampipe. With this granularity of each region, the hit occupancy is expected to be roughly the same over the four regions.

There are two types of chambers in the muon system: one type is the Multi-Wire Proportional Chamber (MWPC); the other type is the triple-Gas Electron Multiplier (GEM). MWPCs are used in the regions in M2-M4 and in the regions R2-R4 of M1. In the region R1 of M1 station, triple-GEM is used. This is due the reason that the innermost region of the M1 station is subjected to a very high particle flux.

Chapter 4

Event processing at LHCb

After pp collisions, there are hits left in the LHCb detectors. One pp collision is called an event. In order to analyze what happened in that event, a process called event reconstruction is performed. It contains several steps: the step to reconstruct tracks from hits in the detector and determine track parameters; the step to find the primary vertex of these tracks; the step to match these tracks to particle identification objects. The frequency of the proton-proton collisions is about 40 MHz. As there are some limitations in data storage, only a portion of collisions is recorded. A trigger system is used to determine whether an event is written to disks or not. After the event reconstruction, only interesting events are selected with stripping lines. The event reconstruction, collision data analysis as well as simulation study are based on LHCb software applications.

4.1 Track reconstruction

The overall performance of the LHCb detector depends on the precise reconstruction of particle tracks. The tracks are formed by combining hits in the VELO, TT, IT and OT detectors. The first step in the track reconstruction is to find all tracks in one event by track pattern recognition algorithms. The goal of the algorithms is to associate correct hits to tracks. In addition, it should minimize the number of reconstructed ghost tracks. A ghost track refers to a fake track reconstructed with randomly selected hits. These random hits are either from hits belong to different tracks or from noise in the detector. The second step is to determine track parameters by a track-fitting algorithm.

4.1.1 Track types

There are five types of tracks in LHCb: long tracks, upstream tracks, downstream tracks, VELO tracks and T tracks[63]. These tracks are classified according to their trajectories in the LHCb spectrometers. They are schematically illustrated in Figure. 4.1. The followings are descriptions of these tracks.

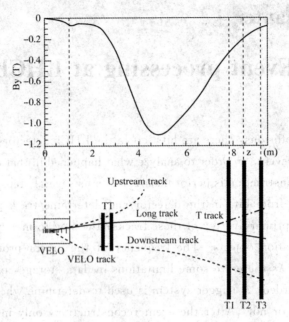

Figure 4.1 A schematic illustration of different track types. The main component of the magnetic field B_y as a function of the z coordinate is also shown on the top of the plot. Taken from Ref.[49]

- Long tracks. Long tracks traverse full tracking systems from the VELO up to the T stations, thus they have the most precise momentum determination. There are two strategies to reconstruct long tracks: the forward tracking and track matching. These two strategies will be described in section 4.1.2.
- Upstream tracks. Upstream tracks only traverse the VELO and TT station. They are low momentum tracks and thus are bent out of the detector acceptance by the magnetic field before reaching the T stations. However they traverse the RICH-1 detector and can generate Cherenkov photons if their velocities are above a threshold. They are therefore used in the RICH-1 reconstruction.
- Downstream tracks. Downstream tracks only traverse the TT and T stations. There are no hits in the VELO for downstream tracks. They reconstruct particles such as K_S^0 which decays outside the VELO acceptance.
- VELO tracks. VELO tracks only traverse the VELO. They are typically large angle tracks. The VELO tracks are used for the primary vertex reconstruction. They can also be used in the identification of backward tracks.

- T tracks. T tracks only traverse the T stations. They are usually produced in secondary interactions. The T tracks are useful in the RICH-2 reconstruction.

Only long tracks are used in the analysis presented in chapter 5. The other types of tracks do not have direct relevance to the analysis.

4.1.2 Track pattern algorithms

The track pattern recognition utilizes six different algorithms to find as many tracks of each type as possible. In the following section the algorithms for the track finding are described.

- VELO seeding. Since the magnetic field in the VELO is very weak (see Figure. 4.1), the VELO tracks are reconstructed as straight lines. The VELO seeding algorithm[66] starts by combining two-dimensional space points in the (r-z) plane into trajectories. These tracks are known as VELO 2D tracks. They can be used to find 2D vertices and to determine 2D impact parameters❶. This information allows a first check in the trigger stage to determine if there are any tracks coming from a secondary decay[67]. A 3D VELO track is reconstructed by adding hits from the VELO ϕ sensors to the 2D VELO track. Both 2D and 3D VELO tracks are reconstructed with hits in at least 3 VELO modules. These VELO tracks act as seeds for the other track finding algorithms.

- Forward tracking. The forward tracking algorithm[68,69] is one of the strategies to reconstruct long tracks. The idea of this algorithm is that the complete trajectory of a particle can be determined by combining the VELO seed track with a single hit in the T stations. The trajectory is parameterized with a parabolic fit in the (x-z) plane and a straight line in the (y-z) plane. Additional hits in the T stations are picked up if they are within a small cone around this trajectory. If hits are picked up in each T station, then a long track is reconstructed. About 90% of long tracks are found with this forward tracking algorithm. In order to reduce the computing time, hits used in the forward tracking algorithm to reconstruct long tracks are discarded in the subsequent track finding algorithms.

- T seeding. The T seeding algorithm is a stand-alone algorithm[70]. It searches for tracks with hits in T stations. These hits are not associated with tracks reconstructed by the forward tracking algorithm. The T tracks are parameterized as a parabola since the magnetic field in the T stations is not negligible.

---

❶ The 2D impact parameter refers to the distance of closest approach between the 2D track and 2D vertex.

- Track matching. The track matching algorithm is another strategy to reconstruct long tracks[71]. It matches T seed tracks with any VELO seed tracks which are not used in the forward tracking algorithm at a plane (indicated as "Plane c" in Figure. 4.2) located behind the last VELO station. In order to do the matching, the T seed track must be extrapolated through the magnetic field to "Plane c". In order to do the extrapolation, the momentum of the T seed track must be known. This momentum is estimated with a p-kick method. In this method, it is assumed that the particle originates from the interaction point and the effect of the magnetic field is approximated by an instant kick of the momentum vector at the center of the magnet. The center of the magnet is in a (x-y) plane (indicated as "Plane a") located at a z point, z_{magnet}, where the integrated magnetic field along the z-axis is a half of the total integrated field. The slope of the T seed at the third station T3 (indicated as "Line 1"), m_1, is extrapolated to "Plane a" and intersects with it at the point $q_{initial}$. A straight line (indicated as "Line 2") joins $q_{initial}$ and the origin. It gives an initial estimate of the track slope inside the VELO, m_2. The momentum of the T seed track then can be evaluated by the difference between m_1 and m_2. This momentum evaluation is shown as follows:

$$\Delta p_x = p \left(\frac{t_{x,f}}{\sqrt{1+t_{x,f}^2+t_{y,f}^2}} - \frac{t_{x,i}}{\sqrt{1+t_{x,i}^2+t_{y,i}^2}} \right) = q \int |\vec{B} \times d\vec{l}|_x \quad (4.1)$$

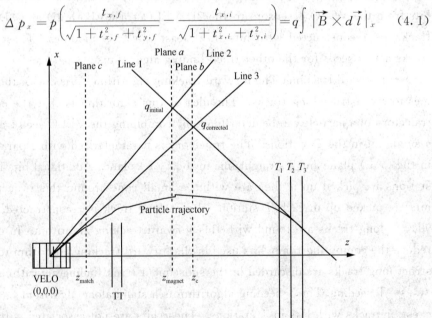

Figure 4.2 A schematic illustration of the p-kick method

where p is the momentum of the T seed track, Δp_x is the change of the momentum in the (x-z) plane, $t_{x,f}$ ($t_{x,i}$) and $t_{y,f}$ ($t_{y,i}$) are slopes of the T seed (VELO seed) track in the x and y directions, $\vec{B}$ is the magnetic field, q is the charge of the T seed track, $d\vec{l}$ is the path defined by the slopes of m_1 and m_2. The precision of this momentum estimation is about 1-2%. The T seed track momentum can be estimated in a more precise way by searching for a point, $q_{corrected}$, on the path. At $q_{corrected}$, the integrated magnetic fields in both directions along the path are equal. A new focal plane (indicated as "Plane b") is defined by the z coordinate of $q_{corrected}$, $z=z_c$. A more precise track slope (indicated as "Line 3") in the VELO, m_3, can be determined by the straight-line connection between $q_{corrected}$ and the origin. The difference between m_3 and m_1 gives a momentum estimation of the track with a precision of about 0.7%[72]. With the momentum information from the p-kick method, the T seed track then is extrapolated through the magnetic field to "Plane c". At "Plane c", the T seed tracks are matched to the VELO seed tracks. In the final step of the track matching procedure, the hits in the TT station are added if they are compatible to the long track.

• Upstream tracking. The upstream tracking algorithm[73] searches for upstream tracks. It collects VELO seeds which remain after the forward tracking or track matching algorithms. These VELO seeds are extrapolated as straight lines in the (y-z) plane to the TT stations. TT hits are searched for in an area that is close to the extrapolation of the VELO seed. These TT hits determine the momentum of the extrapolated track with the p-kick method. An upstream track candidate is formed if at least 3 TT hits give similar momentum estimations. The momentum resolution of upstream tracks is about 15%.

• Downstream tracking. The downstream tracking algorithm[74] searches for downstream tracks. It assumes that the T seeds originate from the interaction point. The momentum of the T seed is estimated with the p-kick method. The candidate T seeds are extrapolated to the TT stations. Compatible hits in the TT stations are added to the extrapolation and downstream track candidates are formed. The momentum resolution of downstream tracks is about 1%.

4.1.3 Track fitting

We have found all tracks by the track pattern recognition. Now we are going to determine their track parameters by the track-fitting algorithm.

Track state

Before describing the track-fitting algorithm, let's first describe the track state. The track state is a set of parameters, which fully describe the position and tangent direction of a particle trajectory at a given z coordinate. Usually a track state is written as

$$\vec{x} = \begin{pmatrix} x \\ y \\ t_x \\ t_y \\ q/p \end{pmatrix} \quad (4.2)$$

where $t_x = \dfrac{\partial x}{\partial z}$ and $t_y = \dfrac{\partial y}{\partial z}$ are track slopes, q is the charge of the particle, $q = \pm 1$. p is the momentum measured from the curvature of the particle trajectory in the magnetic field. The track states can be determined anywhere along the trajectory. In the track fit, they are usually chosen at the planes where the tracking system makes its measurements.

Kalman filter fit

In LHCb, a Kalman filter is applied in the track fit. It adds the measurements on the tracks one by one to the fit. Mathematically the Kalman filter fit is the same as a least squares fit as it is based on minimizing the χ^2 of the measurements on the track. There are several advantages of the Kalman filter fit. First, measurements can be added to the track based on their contributions to the χ^2. There is no need to refit the whole track. Second, the multiple scattering effect and the energy loss which influence the track trajectory are properly accounted in the fit. Third, for tracks with many measurements, the iterative Kalman filter procedure avoids the computation of large dimensional matrix inversions, which are common in the least squares fit. This saves a lot of time and thus makes the fit as fast as possible.

The Kalman filter fit is initialized with an initial track state, $\vec{x}_0$. In the fit the combination of a measurement and a track state is referred to as a node. The fit proceeds iteratively from one node $k-1$ to another node k with the following propagation relation:

$$\vec{x}_k = f_k(\vec{x}_{k-1}) + \vec{w}_k \quad (4.3)$$

where f_k is the function of the track propagation, $\vec{w}_k$ is the process noise such as the multiple scattering. The distance between the measurement, m_k, at the node k and the track state in the measurement plane, $h_k(\vec{x}_k)$ [1], contributes to the χ^2

---

[1] h_k is a projection function. For example, when a detector only measures the x coordinate of a track state, $h_k(\vec{x}_k)$ simplifies to $h_k(\vec{x}_k) = H_k \vec{x}_k$, where H_k is the measurement matrix, $H_k = (1, 0, 0, 0, 0)$.

of the measurements[69]. The purpose of the Kalman fit is to search for optimal track states that give a minimal χ^2. There are three sub-algorithms in the Kalman filter fit procedure:

- Prediction. This prediction algorithm predicts the track state, $\vec{x}_k$, at the node k from the track state, $\vec{x}_{k-1}$, at the previous node $k-1$ with the propagation relation in Eq. (4.3).
- Filter. This filter algorithm updates the predicted track state $\vec{x}_k$ with the measurement information at the node k. The iteration of the prediction and filter steps will not stop until all measurements are added to the track. The track state at the last node, $k=n$, is the best estimate of this track state as all the information from other nodes are included in it. The track states at other nodes, $k<n$, are updated further with the smoother algorithm.
- Smoothing. The smoother algorithm updates the track states at the previous nodes, $k<n$, by reversing the fit iteration from the last node, $k=n$, to the first node. This algorithm makes sure that all the measurements are properly accounted in every node.

4.2 Primary vertex reconstruction

The primary vertex reconstruction is essential to measure impact parameters of particles. There are three primary vertex (PV) finding algorithms. The first algorithm is based on 2D VELO tracks. It is used in the HLT1 trigger stage. The second algorithm is based on 3D VELO tracks. It is used in the HLT2 trigger stage. The third one is based on the best tracks❶. It is used in the offline analysis. For the first two algorithms, they are similar to the "VELO Seeding" algorithm in section 4.1.2 and thus we are not going to describe them in any more detail. In this section, the offline primary vertex reconstruction will be described. There are two steps for the offline primary vertex reconstruction algorithm. In the first step it searches for PV seeds. These seeds will provide the z coordinate information for the PV candidates. In the second step, it reconstructs the PV based on the minimization of the χ^2 in the primary vertex fitting.

---

❶ If two tracks or segments of tracks are duplicates of each other, then a clone killer is applied and only the "better" track is selected. If there are several tracks duplicating to each other, then only the "best" track is selected.

4.2.1 Primary vertex seeding

The PV seeding algorithm employs the method of analytical clusterization[75]. A cluster is a set of one-dimensional coordinates, z^{clu}, and uncertainties associated to the coordinates, σ_z^{clu}. The algorithm proceeds with initial clusters. These initial clusters are intersections of extrapolated tracks with the z axis. The algorithm then merges these clusters iteratively. A pair of clusters will be merged into one cluster if their distance, D, is less than 5. The distance is defined as

$$D = \frac{|z^{clu1} - z^{clu2}|}{\sqrt{(\sigma_z^{clu1})^2 + (\sigma_z^{clu2})^2}} \quad (4.4)$$

The z^{clu} of the merged cluster is determined by the weighted mean of the z^{clu} for the two initial clusters and it is written as

$$z_{merged}^{clu} = \frac{w_1 z^{clu1} + w_2 z^{clu2}}{w_1 + w_2} \quad (4.5)$$

where w_1 and w_2 are weights for the two initial clusters and they are given by $w_i = (1/\sigma_z^{clui})^2$, $i=1, 2$. The merging procedure will stop if there are no more pairs of clusters to be merged. The outputs of the merging procedure are clusters that can not be merged any more. These final clusters serve as primary vertex seed candidates. Finally the algorithm checks qualities of these candidates. If the multiplicity of the cluster is less than 6, it will be removed. This multiplicity is defined as the number of tracks which the cluster consists of.

4.2.2 Primary vertex fitting

The primary vertex fitting algorithm utilizes the long, VELO and upstream tracks in the fitting. Tracks are selected by the requirement that the distance of closest approach between the PV seed candidate position and the track should be less than 30 mm. The algorithm determines the primary vertex position by minimizing the χ^2 of a global least squares fit. The χ^2 is defined as

$$\chi^2 = \sum_{i=1}^{N_{tracks}} \frac{d_{0i}^2}{\sigma_{d_{0i}}^2} \quad (4.6)$$

where d_{0i} is the distance of closest approach between a track and the PV candidate position, and is also referred to as the impact parameter of that track. $\sigma_{d_{0i}}$ is the uncertainty assigned to the distance d_{0i}. The position of the PV is determined iteratively. In each iteration, when the PV position is determined, the impact parameter significance, IPS$=d_0/\sigma_{d_0}$, for each track in the PV determination is estimated. If the IPS of a track is greater than 4, then that track is removed and the new PV position is estimated with the least squares method. This procedure is repeated until no more tracks are discarded. When the final PV position is determined for one primary vertex seed, the primary vertex fitting procedure starts for the next PV seed. In the final step, the qualities of the PV candidates are investigated. Only the PVs with multiplicities above 6 are kept.

Figure. 4.3 shows that the PV resolution in the y direction is a function of the track multiplicity. For a PV composed of 25 tracks in data, the resolution in the (x, y, z) direction is (13.0, 12.5, 68.5) μm^{76}.

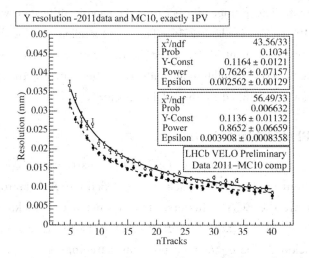

Figure 4.3 Primary vertex resolution is a function of the track multiplicity. The solid (dash) curve shows the resolution result from data (simulation). Taken from Ref.[76]

4.3 Particle identification

Particles in LHCb are identified with the combined information from the RICH detectors, the calorimeter system and the muon system. Electrons are primarily identified by the calorimeter system. Muons are identified by the muon system. Hadrons are identified with two RICH detectors and the hadronic calorimeter. Photons are identified by the electromagnetic calorimeter. Neutral pions, π^0, are detected as two separated clusters or a merged cluster in the electromagnetic calorimeter. The identification techniques for various particles are described in the following sections.

4.3.1 Hadron identification

The RICH system identifies particles with a RICH pattern recognition algorithm[77]. It compares the pattern of hit pixels observed in the RICH photo-detectors with the pattern that is expected under a given set of particle hypotheses. The particle hypotheses are the electron, muon, pion, kaon and proton. A likelihood is calculated based on this comparison. The output of this algorithm is the best hypothesis for each track.

The RICH system does a very good particle identification over the momentum range covered by it. The average efficiency for the kaon identification with the momentum from 2 to 100 GeV/c is about 95%. The corresponding rate for the misidentification of kaons as pions is about 5%[49]. Since the momentum covered by the RICH system is below 100 GeV/c, the high momenta particles generated by the electroweak boson decay❶ can not be identified by the RICH detector[78].

4.3.2 Photon identification

The electromagnetic calorimeter (ECAL) reconstructs and identifies photons as clusters without an associated track[49,79]. A "Cellular Automaton" algorithm[80] is applied to create the ECAL clusters. The reconstructed tracks in an event are extrapolated to the ECAL face. In order to match the ECAL clusters with the reconstructed tracks, a cluster-to-track position-matching estimator, χ_γ^2, is calculated. The minimized χ_γ^2 gives the proximity of the closest track extrapolation and the considered cluster[81]. The clusters due to the charged tracks have a χ_γ^2 spectrum peaking at small values. Photons are identified as clusters with χ_γ^2 greater than 4.

There are two types of photons that can be identified: the converted and unconverted[82]. The converted photon decays into an electron and a positron. The source of the photon conversion comes from the material before the calorimeter. If the conversion takes place before the magnet, the photon will produce two reconstructible electron tracks. However the minimum χ_γ^2 cut on the clusters will remove those converted photons. If the conversion takes place after the magnet, the electrons from the photon are usually not reconstructed and they will make a single cluster in the ECAL. The identification of photons converted after the

---

❶ The typical momentum of the muon from the W boson decay is around 300 GeV/c.

magnet is based on whether a hit exists in the SPD cell that lies in front of the central cell of the ECAL cluster[49].

4.3.3 Electron identification

There are several ways to identify the electrons. They are described as follows.

Identification with the ECAL

The ECAL plays an important role in the electron identification. The electron identification in the ECAL is mainly based on the estimator, χ_e^2, which is built from a global matching procedure between reconstructed tracks and charged clusters in the ECAL[83]. The global matching procedure contains two steps: one step is the balance between the track momentum and energy of the charged cluster in the ECAL; the other step is the matching between the corrected barycenter position of the charged cluster and the extrapolated track impact point. Figure. 4.4 shows the minimum χ_e^2 distributions of electrons (the open histogram) and hadrons (the hatched histogram) in data.

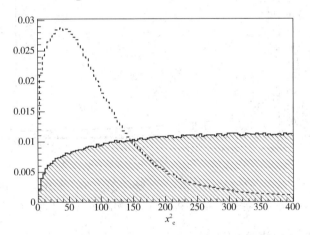

Figure 4.4 The minimum value of the χ_e^2 estimator for the track-cluster energy/position matching procedure. Taken from Ref.[84]

Identification with the PRS detector

The electron identification can be further improved by the utilization of the track energy deposition in the Pre-Shower (PRS) detector. As described in section 3.2.5, the scintillator of the PRS detector is located after the lead absorber. The lead absorber is thick enough for the electron to initiate an electromagnetic showering but insufficient to cause a significant hadronic showering. Figure. 4.5

(a) shows the distributions of energy deposited in PRS for electrons (the open histogram) and hadrons (the hatched histogram) in data. The ionization energy of hadrons deposited in the PRS peaks around 3 MeV[83].

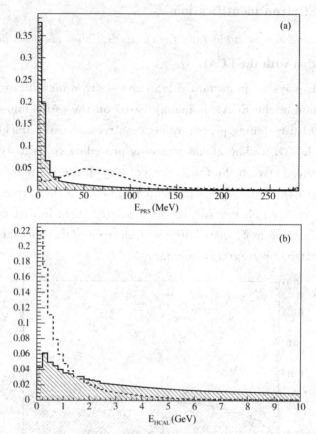

Figure 4.5 (a) The distributions of the energy deposited in the PRS.
(b) The distributions of the energy deposited in the HCAL. Taken from Ref.[84]

Identification with the HCAL

The energy deposited in the HCAL can also be used to identify electrons. Since the total thickness of the ECAL corresponds to a radiation length of 25 X_0, there is a very small leakage of the electromagnetic shower into the HCAL. Figure. 4.5(b) shows the distributions of the energy deposited in the HCAL for electrons (the open histogram) and hadrons (the hatched histogram) in data.

Identification with Bremsstrahlung Photons

Photons are emitted from electrons in the region before or after a sizable magnetic field. As there is little material in the magnet, if the photon is emitted before the magnet, its position can be predicted by a simple linear extrapolation of the reconstructed track segment before magnet to the ECAL plane (see Figure. 4.6). If the photon is emitted after the magnet, it will form a part of the electron cluster. Only the bremsstrahlung photons emitted before the magnet are utilized to identify electrons. For the photon emitted before the magnet, a χ^2_{brem} is constructed between the predicted position of the bremsstrahlung photon and the barycenter position of identified photons in section 4.3.2. The minimum value of χ^2_{brem} is utilized as a discrimination variable in the electron identification.

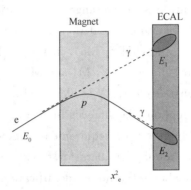

Figure 4.6 Schematic illustration of bremsstrahlung photons. Taken from Ref.[63]

High momentum electron identification

As described in section 2.5, there is saturation in the energy readout of the cell in ECAL, the E_T threshold of this saturation is around 10 GeV. As a result, the identification of the high momentum electron in LHCb is a challenge. The details of the high momentum electron identification can be found in Ref. [78].

4.3.4 Muon identification

There are two muon identification procedures in LHCb: one is the online muon identification which is vital in the L0 trigger stage; the other one is the offline muon identification. Since the physics channel presented in this book contains one muon in the final state, the method to identify the offline muons will be described in detail in the following section. In the offline muon identification algorithm, well-reconstructed tracks whose momenta are greater than 3 GeV/c are extrapolated into muon stations. These tracks must lie in the acceptance of muon stations from M2 to M5. Then the algorithm searches for muon detector hits around a field of interest (FOI). The FOI is built around the extrapolation point of the reconstructed track in each muon station. The size of the FOI is different in the x and y directions and it is parameterized as a function of the track momentum, p, for each station and each region[85]. The function is written as

$$\text{FOI} = p_0 + p_1 \cdot \exp(-p_2 \cdot p) \qquad (4.7)$$

where p_0, p_1 and p_2 are parameters and are chosen from Monte Carlo to maximize the muon identification efficiency and to maintain a low level of pion misidentification[63]. A reconstructed track with $3 < p < 6$ GeV/c is considered to be a muon candidate if hits are found around the FOIs in the muon stations M2 and M3. If the track's momentum is in the range of $6 < p < 10$ GeV/c, it is identified as a muon candidate with the requirement that hits are found around the FOIs in the muon stations M2, M3 and (M4 or M5). If the track's momentum is above 10 GeV/c, it must have hits in M2, M3, M4 and M5. A boolean value, IsMuon, is set to be true if a reconstructed track is identified as a muon candidate. This boolean value is used throughout the analysis in chapter 5.

The efficiency of IsMuon = True requirement is defined as the efficiency of finding hits within FOI in muon chambers for tracks extrapolated to the muon system[86]. Figure. 4.7(a) shows the muon identification efficiency in data, ε_{IM}, as a function of the muon momentum for different transverse momentum ranges. This efficiency is weakly dependent on the muon transverse momentum. It drops about 2% in the first p_T interval. The reason for this drop is explained as follows: tracks close to inner edges of the region R1 are scattered outside the detector, thus extrapolation points of these tracks are not within the M1 and M5 acceptance. In the momentum range from 3 to 100 GeV/c, the muon identification efficiency for particles with $p_T > 1.7$ GeV/c is above 97%. Figure. 4.7(b) (Figure. 4.7(c)) shows the probability of pions (kaons) misidentified as muons in data, $p_{IM}(\pi \to \mu)$ ($p_{IM}(K \to \mu)$), for different transverse momentum ranges. As tracks with higher p_T traverse the detector at higher polar angles and in lower occupancy regions, the misidentification probability decreases when the transverse momentum increases. The average pion (kaon) misidentification probability is $(1.033 \pm 0.003)\%$ $((1.025 \pm 0.003)\%)$. In the analysis presented in chapter 5, the muon identification efficiency is evaluated with a requirement of $p_T > 20$ GeV/c.

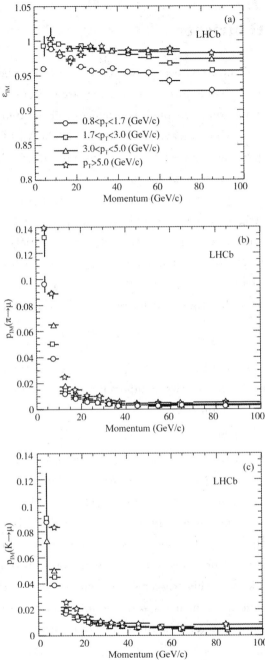

Figure 4.7 (a), muon identification efficiencies as a function of particle momentum. (b), pion misidentification probabilities as a function of particles momenta. (c), kaon mis-identification probabilities as a function of particle momentum. The circle (square, triangle, star) dots are for $0.8 < p_T < 1.7$ ($1.7 < p_T < 3.0$, $3.0 < p_T < 5.0$, $p_T > 5.0$) GeV/c. Taken from Ref.[86]

4.4 LHC btrigger

The frequency of the LHC beam crossing is 40 MHz. About 10 MHz of the beam crossings contain visible interactions which are detected by the LHCb spectrometers[49]. An interaction is visible if it produces at least two charged particles. These charged particles should leave sufficient hits in the VELO and T stations in order to allow the particles' trajectories to be reconstructed. A trigger system in LHCb is utilized before these huge amount of events are written into disk. It reduces the rate of events to 2 KHz. This reduction is done by a sequence of trigger stages: the level 0 trigger (L0); the high level trigger 1 (HLT1); the high level trigger 2 (HLT2). In each trigger stage, only events satisfying trigger requirements are recorded. In order to maintain a high efficiency for signal events and reject background events, the requirements in the trigger stages are optimized. In the following sections, the trigger stages of L0, HLT1 and HLT2 are described.

4.4.1 L0

The L0 trigger is a hardware level trigger. It reduces the rate of events from 10 MHz to 1 MHz. There are two components in the L0 trigger stage: the L0 calorimeter trigger and muon trigger.

L0 calorimeter trigger

The purpose of the L0 calorimeter trigger is to look for high transverse energy (E_T) particles[49]. These particles are electrons, photons, π^0 or hadrons. The alorimeter trigger uses a zone of 2×2 cells as a cluster. This zone is large enough to contain most of the shower energy of a particle while it is small enough to avoid an overlap of the shower energy depositions for various particles. In order to minimize the number of candidates to be processed, only clusters with the largest E_T are selected and kept at this trigger stage. These selected clusters are identified as electrons, photons or hadrons with the information from the SPD, PRS, ECAL and HCAL calorimeters. The thresholds of E_T values for different types of particles are listed in Table 4.1. The total E_T in all HCAL cells is a criterion to reject crossings without visible interactions. The total number of SPD cells with a hit is called the SPD multiplicity. Events with a large SPD multiplicity contain a large number of final state particles and will spend a lot of processing time. These events can be rejected by applying a SPD multiplicity cut. As the SPD multiplicity is a global feature of collision events, this multiplicity cut is applied with the same requirements to all L0 trigger stages.

Table 4.1 **L0 transverse energy thresholds for different types of particles. Taken from Ref.[72]**

Particle Type	Hadron	Electron	Photon	π^0
E_T threshold (GeV)	3.6	2.8	2.6	4.5

L0 muon trigger

The purpose of the L0 muon trigger is to look for muon tracks with large transverse momenta[88]. The muon tracks are reconstructed with a track finding algorithm. This algorithm uses hits in the muon station M3 as seeds. Then it joins the hit in M3 and the interaction point as a straight line. This line is extrapolated to other muon stations and hits are searched for in the FOIs centered on extrapolation positions in M2, M4 and M5. The size of the FOI depends on the muon station considered and the background level allowed[49]. If there is at least one hit inside the FOI for each station of M2, M4 and M5, then the track is identified as a muon. After hits are found in the muon stations of M2, M4, and M5, hits in M1 are searched for. This is done by a straight-line extrapolation from hits in M3 and M2 to M1. A hit then is looked for in a FOI around the extrapolation point in M1. At last the algorithm will utilize the hits in M1 and M2 to measure the p_T of the muon track. As the effect of the magnetic field is not negligible, the muon p_T measurement is done by the p-kick method. The resolution of the muon track p_T is about 20%. The threshold on the muon p_T for the single muon trigger is 1.36 GeV/c. For the dimuon trigger the requirement is that the sum of the transverse momenta of the two muons should be greater than 1.48 GeV/c[87]. This means that each of these two muons could have a transverse momentum below the single muon trigger threshold of 1.36 GeV/c. In the analysis presented in the book, only the L0 single muon trigger is utilized.

4.4.2 HLT1

The HLT1 is a software level trigger[49]. It is the first step of the high level trigger. It runs on events passing the L0 trigger. These events are called L0 objects. The input rate of the HLT1 is about 1 MHz while the output rate is about 30 MHz. The HLT1 commences with alleys (see Figure. 4.8). An alley is a set of algorithms[88]. Each alley corresponds to one of the L0 objects. The algorithms in alleys firstly confirm L0 objects with the reconstruction of particles in the VE-

L0 and T stations. Then they impose specific requirements on the reconstructed objects in order to pass them. These requirements are organized into HLT1 trigger lines. If a muon candidate satisfies requirements in a trigger line, then we say this trigger line is fired.

For muons, there are two alleys in the HLT1 trigger stage: a single muon alley, where a p_T cut is applied on a single muon, and a dimuon alley, where p_T cuts are imposed on the two highest p_T muons. Cuts are also applied in other alleys such as hadrons, electrons and neutrals (e.g. π^0 and γ).

4.4.3 HLT2

The HLT2 is another software trigger stage of the high level trigger. The input rate of the HLT2 is about 30 MHz while the output rate is about 2 KHz. Since the input rate of events for the HLT2 is relatively low, it allows the HLT2 to perform a full offline track reconstruction. The procedure to reconstruct muons in the HLT2 trigger stage is similar to the offline muon reconstruction described in section 4.3.4. Due to time constraints, the reconstructed tracks are not fitted with the Kalman filter described in section 4.1.3. Therefore it is important to apply some loose track quality cuts on these reconstructed tracks before the offline analysis. Inclusive and exclusive selections are applied in the HLT2 to reduce the rate to 2 KHz. These selections are organized into HLT2 trigger lines. Events passing through a certain HLT2 trigger line are written to storage for further analysis. The flow of all trigger stages at LHCb is provided in Figure. 4.8.

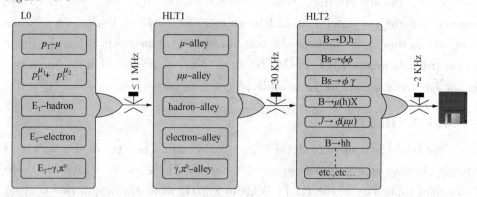

Figure 4.8 The flow of the trigger stages at LHCb. The event rates at each trigger stage are indicated. HLT1 confirms the L0 objects with alleys. In HLT2, different trigger lines are utilized to select events for offline analysis. Taken from Ref.[49]

4.4.4 Global event cuts

Events with extremely large occupancy in the VELO, OT, IT and SPD sub-detectors will spend a lot of processing time. In order to reject these events, Global Event Cuts (GECs) are applied to these sub-detectors variables event by event at L0, HLT1 and HLT2 trigger stages and are detailed in the next section.

4.4.5 LHCb trigger lines

We have described the trigger stages at L0, HLT1 and HLT2. In the following section trigger lines utilized in this book are presented. There were two distinct data-taking periods in 2010 during which LHCb collected 37.1 pb^{-1} of data. In the first stage (stage 1), LHCb collected about 2.8 pb^{-1} of data[91]. In the second stage (stage 2), the rest 34.3 pb^{-1} of data was collected. In different data collection stages, different trigger requirements were utilized.

• Requirements at L0. In the L0Muon trigger line, a kinematic cut in stage 1 required that the p_T of the muon should be greater than 1.25 GeV/c and a GEC in stage 1 required the SPD multiplicity should be smaller than 600. In stage 2, the muon p_T threshold was increased to 1.75 GeV/c and the SPD multiplicity cut was changed to 900. Another trigger line, L0MuonHigh, was also utilized in the L0 trigger stage. In stage 1, the p_T and SPD multiplicity cuts were not presented in this line. However, in stage 2, in order to fire this trigger line, the muon p_T should be greater than 3 GeV/c and the SPD multiplicity should be smaller than 900. If a muon candidate fired any of the two trigger lines, then it was passed to the HLT1 trigger stage.

• Requirements at HLT1. In HLT1, Hlt1SingleMuonNoIPL0 and Hlt1SingleMuonNoIPL0HighPT trigger lines were utilized. In order to fire the first trigger line, the transverse momenta of muons should be greater than 1.35 GeV/c in stage 1 and 1.8 GeV/c in stage 2. The second trigger line did not apply requirements on the muon p_T in stage 1 while in stage 2 it required the muon p_T should be greater than 5 GeV/c. Besides the kinematic cuts applied in the HLT1 trigger stage, the following GECs had been applied: the number of VELO clusters $<$ 3000, the number of clusters in the IT station $<$ 3000, the number of clusters in the OT station $<$ 10000 and the number of VELO tracks $<$ 350. If a muon candidate satisfied the GECs and fired any of the two trigger lines, then it was passed to the HLT2 trigger stage.

- Requirements at HLT2. The Hlt2SingleHighPTMuon trigger line required that the p_T of the muon should be greater than 10 GeV/c in both stage 1 and stage 2. In the HLT2 trigger stage, no global event cuts were applied. If a muon candidate fired this trigger line, then it was kept for the offline analysis.

4.5 LHCb stripping

The LHCb stripping is a centralized event selection after the event reconstruction. It selects interesting events with a stripping line. In the stripping line there are cuts applied to create candidates and select events of specific interest. The stripping lines used in this analysis are presented as follows:

- W2Mu line. The W2Mu line is used to get a data sample with muon candidates from W bosons. In this line, candidate tracks are identified as muons and their transverse momenta are greater than 20 GeV/c.
- W2MuNoPIDs line. The W2MuNoPIDs line is used to get a decay in flight data driven sample. In this line, no identification information is associated with candidate tracks. The transverse momenta of these tracks are greater than 15 GeV/c.
- MBNoBias line. The MBNoBias line is used to get a minimum bias sample. In this line, candidate tracks are randomly triggered and they are unbiased to any given process. No identification information is associated with these tracks.
- Z02MuMu line. The Z02MuMu line is utilized to get a Pseudo-W data sample. In this line, candidate tracks are identified as muons and their transverse momenta are greater than 15 GeV/c. The invariant mass of the dimuon pair must be greater than 40 GeV/c^2 in this line.
- Z02MuMuNoPIDs line. The Z02MuMuNoPIDs line is used to get a $Z \to \mu\mu$ data sample. This data sample is used to calculate the muon track identification efficiency. In this line, no identification information is associated with candidate tracks. The transverse momenta of these tracks are greater than 15 GeV/c.

4.6 LHCb software

The LHCb software is built within an object orientated framework[92]. This framework is known as Gaudi. It supports several data processing software applications. These applications include simulation, digitization, reconstruction and physics analysis. The simulation is done by the Gauss package[93]. The digitization is done by the Boole package[94]. The reconstruction is done by the Brunel package[95]. The physics analysis is done by the DaVinci package[96]. These packages are described in the following sections.

4.6.1 Gauss package

The Gauss package mimics the event generation in the proton proton collision at LHCb. It allows understanding the LHCb experimental conditions and performance. There are two independent phases during the Gauss simulation. The first phase is a generator phase. It consists of the event generation in the pp collisions and the decay of the unstable particles produced. The second phase is a simulation phase. It simulates the passing of the generated particles through the LHCb detector and the interaction between the particles and the detector.

For the generator phase, it utilizes PYTHIA 6.4[36] as a general purpose generator for pp collisions. It also interfaces with an external generator library, LHAPDF[97], to provide parton distribution functions. The default parton distribution function used in PYTHIA 6.4 is CTEQ5L[98].

For the simulation phase, it utilizes GEANT4[99] to simulate the effects of the multiple scattering, energy loss and the photon conversions. In order to make the simulation accurate, the LHCb detector is simulated in detail. Active detector elements and passive materials such as the support structures, frames, shielding elements and the LHCb beam pipe are fully described with GEANT4. The secondary interactions of hadrons with momentum greater than 10 MeV/c can be fully simulated. For the interactions of leptons and photons, the momentum threshold for the simulation is 1 MeV/c. The impact of the magnetic field in LHCb is simulated with a field map which was measured after the magnet is installed.

4.6.2 Boole package

The Boole package simulates the response of the LHCb detector to the simulated physics events from Gauss. This simulation performs the digitization of the LHCb detector. It is the final stage of the LHCb detector simulation. The digitization step includes two simulations: one is the simulation of the response of each sub-detectors to the hits previously generated in sensitive detectors simulated by GEANT4; the other one is the simulation of the detector response to the hits from spillover events and LHCb backgrounds. Test beam data has been used to tune the simulation of the sub-detector response to the hits. The L0 trigger hardware is also simulated. The output of the digitized data from the Boole package is in the same format as the format of the data coming from the real detector.

4.6.3　Brunel package

The Brunel package is the LHCb event reconstruction application. It processes digitized events either from data, or from the output of the detector digitization application, Boole. There are several algorithms in the Brunel package. These algorithms utilize a full pattern recognition on the sub-detector hits to form tracks and vertices. They also use particle identification algorithms on the information from the calorimeters, RICH detectors and the muon systems to identify particles.

4.6.4　DaVinci package

The DaVinci package is the physics analysis software for the LHCb experiment. It analyses events produced by the Brunel package. Tracks and vertices can be refitted in DaVinci. Resonances can be formed by the combination of particles. Offline selection algorithms can be developed and applied in DaVinci to select interesting physics processes as well as to reject background events.

Chapter 5

$\sigma_{W \to \mu \nu_\mu}$ measurement at LHCb

By measuring the $W \to \mu \nu_\mu$ cross-section at LHCb, the electroweak theory can be tested in a new energy regime. In this chapter, the analysis of W production at $\sqrt{s} = 7$ TeV with 37.1 ± 1.3 pb^{-1} of data collected by the LHCb experiment during 2010 is presented. In order to select candidate tracks with good quality, some track pre-selection cuts are applied. As $W \to \mu \nu_\mu$ signal events are produced with the accompaniment of background events, some candidate selection cuts are applied to suppress these background events. In order to optimize these cuts, a Pseudo-W data sample is utilized. A fit is performed to determine the purity of signal events in data. After the fit, the purity is found to be about 79%. Due to the trigger procedure of the experiment, the imperfect performance of the track reconstruction and particle identification, the selection cuts, and the fiducial cuts with the muon p_T between $20 < p_T^\mu < 70$ GeV/c, the number of signal events are underestimated. This number is corrected with a factor that is the inverse of efficiencies due to these cuts. Finally, $W^+ \to \mu^+ \nu_\mu$ and $W^- \to \mu^- \bar{\nu}_\mu$ cross-sections in a fiducial region with $p_T^\mu > 20$ GeV/c are calculated and the $W \to \mu \nu_\mu$ cross-section ratio and charge asymmetry are shown. The W cross-sections, cross-section ratio and charge asymmetry are compared to theoretical predictions. The W cross-sections and their charge asymmetry are also compared to the results measured at ATLAS.

5.1 Signal and background processes

The signal in the analysis presented in the book is the muon coming from the W boson decay. The Feynman diagram for the W production and decay is shown in Figure. 5.1. q_i is from one proton, $\bar{q}_j$ is from another proton. They annihilate to produce a W boson. The W boson then decays into a muon and a neutrino.

A number of background processes look similar to the signal and they are considered in this analysis.

The first background is from kaons or pions which decay in flight into muons and neutrinos. These muons are detected in the muon chambers. Alternatively, kaons or pions punch through into muon chambers if their energies are large enough. These kaons or pions can be reconstructed as muons.

The second background considered is from mesons containing b ($\bar{b}$) quarks in the heavy flavor process. A b quark decays into a c or u quark and a W boson. The W boson then decays into a muon and neutrino. The Feynman diagram for the b quark decay is illustrated in Figure. 5. 2.

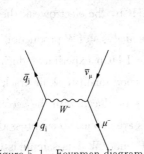
Figure 5.1 Feynman diagram for the W production and decay

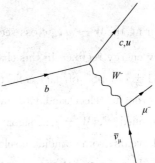
Figure 5.2 Feynman diagram for the b quark decay

The third background is from mesons containing c ($\bar{c}$) quarks in the heavy flavor process. A c quark decays into a d or s quark and a W boson. The W boson then decays into a muon and neutrino. The Feynman diagram for the c quark decay is illustrated in Figure. 5. 3.

The fourth background is the $W \to \tau \nu_\tau$ process. The Feynman diagram for $W^- \to \tau^- \bar{\nu}_\tau \to \mu^- \bar{\nu}_\tau \bar{\nu}_\mu \bar{\nu}_\tau$ is shown in Figure. 5. 4. q_i is from one proton, $\bar{q}_j$ is from another proton. They produce a W boson. The W boson then decays into a tau lepton and a tau neutrino. The tau lepton subsequently decays into a muon lepton, a muon neutrino and a tau neutrino.

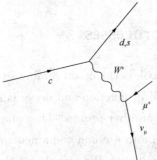
Figure 5.3 Feynman diagram for the c quark decay

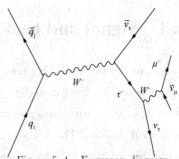
Figure 5.4 Feynman diagram for the $W \to \tau \nu_\tau$ process

The fifth background is the $Z \to \tau\tau$ process. The Feynman diagram for $Z \to \tau^+ \tau^-$ is shown in Figure. 5.5. q_i is from one proton, $\bar{q}_i$ is from another proton. They produce a Z boson which decays into a τ^+ and τ^-. τ can decay into a μ, e or a hadron. There are two cases for τ decays. One case is that one τ decays into a muon in the LHCb acceptance, the other τ decays into an electron or a hadron inside or outside the LHCb acceptance. The other case is that both τ decay into muons in the LHCb acceptance. As the second case can be removed by requiring only one muon in the LHCb acceptance, only the first case is considered as a background for the $W \to \mu\nu_\mu$ analysis.

The last background considered is from the $Z \to \mu\mu$ process. The Feynman diagram for the Z boson production and decay is shown in Figure. 5.6. q_i is from one proton, $\bar{q}_i$ is from another proton. They produce a Z boson which then decays into μ^+ and μ^-. Muons can be inside or outside the LHCb acceptance. We only consider the case in which only one reconstructed muon is in the LHCb acceptance as a background for the $W \to \mu\nu_\mu$ analysis.

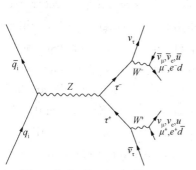

Figure 5.5 Feynman diagram for the $Z \to \tau\tau$ process

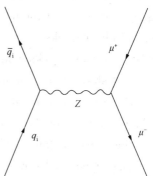

Figure 5.6 Feynman diagram for the $Z \to \mu\mu$ process

5.2 Track pre-selection requirements

In order to select tracks with good quality for the analysis, some track pre-selection requirements are applied on events from the W2Mu stripping line. The quality of a track fit is reflected by the fit χ^2 and relative momentum resolution. As described in section 4.1, ghost tracks are reconstructed with random combinations of hits in the tracking system. Therefore these ghost tracks do not follow trajectories of real tracks and are characterized by poor quality of track fits. Ghost tracks can be suppressed by a requirement that the number of hits in the TT station, N_{TThits}, for a track should be larger than 0. Finally, in order to minimize detector edge effects, the pseudo-rapidity of a track, η_{track}, should be in the range of $2 < \eta_{track} < 4.5$.

5.2.1 Track χ^2 probability requirement

In a simulation sample, if a reconstructed track shares less than 70% of its detector hits with a generator level particle, it is deemed as a ghost track[89]. However, this method to identify tracks as ghost tracks in the simulation sample does not work 100% correctly as ghost tracks occasionally share more than 70% of their hits with generator level particles. Figure. 5.7(a) shows χ^2 probability distributions for ghost tracks (solid line) and non-ghost tracks (dash line) with momenta greater than 10 GeV/c in the simulation sample. The χ^2 probability distribution for ghost tracks peaks below 0.01. This is due to bad reconstructions of these ghost tracks. As most of non-ghost tracks are well reconstructed, their χ^2 probability distribution is almost flat except for a small peak below 0.01, which represents about 8% of the non-ghost sample, and is due to poor reconstructions or residual ghost tracks. A cut with χ^2 probabilities greater than 1% will remove about 70% of ghost tracks while it will keep about 92% of non-ghost tracks. Figure. 5.7(b) shows χ^2 probability distributions for tracks with momentum greater than 10 GeV/c in the simulation and data samples. As some tracks are badly reconstructed both in the data and simulation samples, the χ^2 probability distributions of these two samples peak below 0.01. The χ^2 probability distributions for well-reconstructed tracks in both samples are flattish between 0 and 1. However the relative amounts of well-reconstructed tracks and badly reconstructed tracks are different in these two samples. This is due to the reason that hit resolutions in data are not well simulated. As the χ^2 probability distribution in the simulation sample is different from the distribution in the data sample, we only use the data information to work out detector efficiencies. In order to remove badly reconstructed tracks both in the data and simulation samples, a cut with χ^2 probabilities larger than 1% is applied.

5.2.2 TThits requirement

As described in section 3.2.3, the TT station is in front of the magnet and just behind the RICH1. The pseudo-rapidity range of the TT station, η_{TT}, is $2 \lesssim \eta_{TT} \lesssim 4.5$. The pseudo-rapidity range of LHCb, η_{LHCb}, is $1.9 < \eta_{LHCb} < 4.9$.

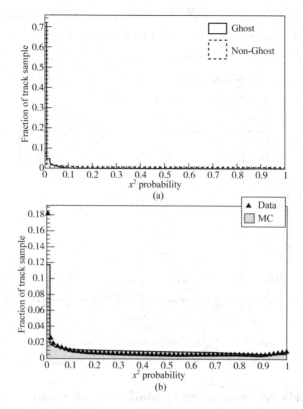

Figure 5.7 (a), χ^2 probability distributions for ghost tracks and non-ghost tracks in the simulation sample. (b), χ^2 probability distributions for tracks in the simulation and data samples

As described in section 4.1.2, some tracks are reconstructed with the track-matching algorithm. This algorithm tries to match the T seed track to any VELO seed tracks. A matching χ^2 is calculated according to the track parameters of the VELO seed track and T seed track. If this χ^2 is less than a predefined cut, then the combination of the VELO seed track and T seed track is deemed as a real (non-ghost) track[71]. If the pseudo-rapidity of that real track is in the η range of the TT station, then there ought to be corresponding TT hits in the TT station. Ghost tracks are reconstructed with random combinations of VELO seed tracks and T seed tracks. If the pseudo-rapidity of a reconstructed track is in the η range of the TT station, and there are no TT hits associated to that reconstructed track, then that reconstructed track is likely to be a ghost track. Due to the fact the pseudo-rapidity range of TT station is smaller than the range of LHCb, a small number of real tracks do not have TT hits❶, but most of the ghost tracks do not have TT hits. Figure. 5.8 shows N_{TThits} distributions for ghost tracks (solid line) and non-ghost tracks (dash line) with momenta greater than 10 GeV/c in the simulation sample. Applying the requirement of $N_{TThits}>0$ can remove about 70% of ghost tracks while it can keep about 80% of real tracks.

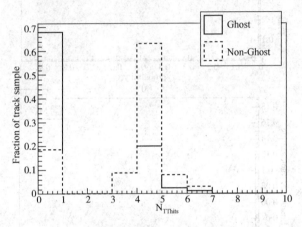

Figure 5.8 N_{TThits} distributions for ghost tracks and non-ghost tracks in the simulation sample

5.2.3   Relative momentum resolution requirement

The relative momentum resolution of a track is defined as σ_p/p, where σ_p is the uncertainty of the track momentum, p is the track momentum. Figure. 5.9 (a) shows the scatter plot of σ_p/p as a function of the track momentum for non-ghost tracks with momenta greater than 10 GeV/c in the simulation sample. A cluster of tracks with momenta smaller than 100 GeV/c has poor resolutions (5% to 10%). This is due to the residual ghost tracks. Most of non-ghost tracks in this distribution are confined in a region with $\sigma_p/p<0.1$. Figure. 5.9(b) shows the scatter plot of σ_p/p as a function of the track momentum for ghost tracks with momenta greater than 10 GeV/c in the simulation sample. Some ghost tracks in this distribution are in a region with $\sigma_p/p>0.1$. Figure. 5.9(c) shows the scatter plot of σ_p/p as a function of the track momentum for all ghost and non-ghost tracks in the data sample. The region with $\sigma_p/p>0.1$ in this distribution is more densely populated than the region with $\sigma_p/p>0.1$ for ghost tracks in the simulation sample. Obviously there are more ghost tracks in the data sample

---

❶   These real track's pseudo-rapidity is greater than 4.5.

than in the simulation sample. This is due to the reason that the detector simulation is imperfect in the simulation sample. From simulation, we know that applying a cut with $\sigma_p/p < 0.1$ keeps about 99.99% of real tracks while it removes about 0.08% of ghost tracks.

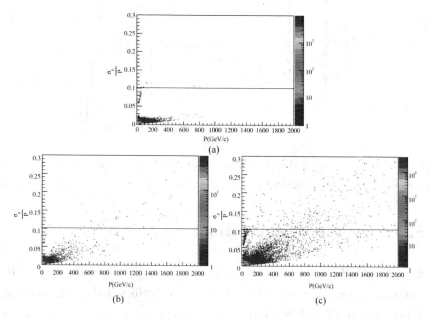

Figure 5.9 (a) ((b)), scatter plot of σ_p/p as a function of the track momentum for non-ghost tracks (ghost tracks) in the simulation sample. (c), scatter plot of σ_p/p as a function of the track momentum for all ghost and non-ghost tracks in the data sample

5.2.4 Track pseudo-rapidity requirement

Figure. 5.10 shows the pseudo-rapidity distributions of truth and reconstructed muons in the $W \to \mu \nu_\mu$ simulation sample. In this figure, the fraction of events for the reconstructed muon distribution is significantly smaller the fraction of events for the truth muon distribution in the ranges $\eta < 2.0$ and $\eta > 4.5$. This could be explained as follows: as it is difficult to reconstruct muons close to the edge of the LHCb detector, there is inefficiency for the muon track reconstruction. In order to minimize this detector edge effect, a requirement that $2.0 < \eta < 4.5$ is applied on the muon tracks in both the simulation and data samples.

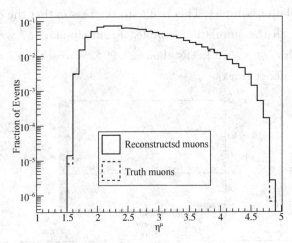

Figure 5.10 Pseudo-rapidity distributions of truth (dash line) and reconstructed (solid line) muons in the $W \to \mu \nu_\mu$ simulation sample. Both distributions are normalized to unit area

5.3 Pseudo-W data sample

The Pseudo-W data sample is a fake sample of the $W \to \mu \nu_\mu$ decay. It contains a dimuon pair consistent with a Z boson in the final state. The dimuon pair is selected from the Z02MuMu stripping line. The presence of a neutrino is mimicked by masking each of muons in the dimuon pair alternately[51]. There are two usages for the Pseudo-W data sample. One is to optimize the W boson selection criteria described in section 5.4. The other one is to calculate the selection efficiency for the W boson described in section 5.8.

The following requirements have been applied on the dimuon pair in order to produce the Pseudo-W data sample:

- At least one muon fires the trigger lines described in section 4.4.5.
- The transverse momentum of each muon is greater than 20 GeV/c.
- A dimuon invariant mass is in the range of $81 \text{ GeV}/c^2 < m_{\mu\mu} < 101 \text{ GeV}/c^2$.

Section 5.4 shows that general agreement is achieved between the Pseudo-W data sample and $W \to \mu \nu_\mu$ simulation sample. However, there is a difference between the p_T spectra in the $W \to \mu \nu_\mu$ and $Z \to \mu\mu$ decays. The p_T spectrum is harder in the $W \to \mu \nu_\mu$ decay than in the $Z \to \mu\mu$ decay. This difference is estimated with the $W \to \mu \nu_\mu$ and $Z \to \mu\mu$ simulation samples. The Pseudo-W data sample then is corrected with this difference.

5.4 Candidate selection cuts

In order to suppress each source of background and isolate events of interest, a set of cuts have been applied to the well reconstructed muon candidates whose transverse momenta are in the range of 20 GeV/c $< p_T^\mu <$ 70 GeV/c.

5.4.1 Extra muon transverse momentum cut

$Z \to \mu\mu$ background events with both muons in the LHCb acceptance and $Z \to \tau\tau$ background events with both $\tau \to \mu\nu_\mu\nu_\tau$ decays in the LHCb acceptance are removed by a requirement that there is only one muon candidate in the event with $p_T^\mu >$ 20 GeV/c and any other muons in the event should have small transverse momenta (p_T^{Extra}).

Figure. 5.11 shows the muon p_T^{Extra} distributions for the Pseudo-W data sample, $W \to \mu\nu_\mu$ and $Z \to \mu\mu$ simulation samples. Points with error bars are for the Pseudo-W data sample, the histogram with the solid (dash) line is for the $W \to \mu\nu_\mu$ ($Z \to \mu\mu$) simulation sample. Good agreement is achieved between p_T^{Extra} distributions of the Pseudo-W data sample and $W \to \mu\nu_\mu$ simulation sample. There are very few candidates with the extra muon p_T greater than 2 GeV/c for these two samples. For the $Z \to \mu\mu$ simulation sample there is an obvious peak around 45 GeV/c. This is due to the reason that there is another high p_T muon from the Z boson. Applying a requirement of $p_T^{Extra} >$ 2GeV/c can remove about 91% of $Z \to \mu\mu$ background events while it can keep about 89% of signal events.

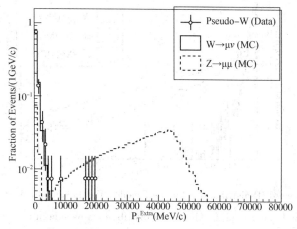

Figure 5.11 Muon p_T^{Extra} distributions for the Pseudo-W data sample, $W \to \mu\nu_\mu$ and $Z \to \mu\mu$ simulation samples. These three distributions have been normalized to unit area

5.4.2 Impact parameter cut

The impact parameter, IP, is defined as the distance of closest approach of the muon candidate track to the primary vertex (PV). Since the lifetime of W bosons is much shorter than those of B-hadrons and C-hadrons[❶], muons from W bosons decay typically have a smaller IP than muons from B-hadron and C-hadron decays. As described in section 4.2, muons from W bosons take a large weight in the primary vertex reconstruction. Thus the primary vertex position is strongly shifted towards the muon candidate track from the W boson decay. In order to get an unbiased primary vertex, the primary vertex is re-fitted by excluding the muon candidate track from the W boson decay. The distance of closest approach of the muon candidate track to the new primary vertex is called the unbiased IP.

Figure. 5.12 shows the unbiased impact parameter distributions for the Pseudo-W data sample, $W \to \mu \nu_\mu$, $b \bar{b} \to X\mu$ and $c \bar{c} \to X\mu$ simulation samples. Points with error bars are for the Pseudo-W data sample, the histogram with the solid (dash, dash−dot) line is for the $W \to \mu \nu_\mu$ ($b \bar{b} \to X\mu$, $c \bar{c} \to X\mu$) simulation sample. Good agreement is achieved between the unbiased impact parameter distributions of the Pseudo-W data sample and $W \to \mu \nu_\mu$ simulation sample. There is a sharp peak around the small impact parameter in both of these two unbiased IP distributions. There are broader tails in the unbiased IP distributions of the $b \bar{b} \to X\mu$ and $c \bar{c} \to X\mu$ simulation samples.

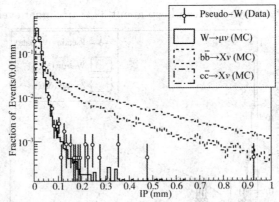

Figure 5.12　Unbiased IP distributions for the Pseudo-W data sample, $W \to \mu \nu_\mu$, $b \bar{b} \to X\mu$ and $c \bar{c} \to X\mu$ simulation samples. These four distributions have been normalized to unit area

---

❶　The lifetime of W boson is around 10^{-25} s, while the lifetime of B meson is around 10^{-12} s.

A cut with IP < 40 μm can remove about 55% (80%) of muons from the $b\bar{b} \to X\mu$ ($c\bar{c} \to X\mu$) background events while it can keep around 93% of signal events.

5.4.3 Relative energy deposition cut

There is very little energy deposited in the ECAL or HCAL when muon candidates travel through these two calorimeters. Kaons or pions deposit more energy than muon candidates when they punch through calorimeters into muon chambers. In order to reject the punch through background, the deposited energy in the ECAL (E_{ECAL}) and HCAL (E_{HCAL}) around the extrapolated muon track are added together as $E_{ECAL+HCAL}$. The relative energy deposition in calorimeters is defined as $E_{ECAL+HCAL}$ divided by the muon track momentum. It is written as $E_{ECAL+HCAL}/P$.

Figure. 5.13 shows the relative energy deposition distributions in calorimeters for the Pseudo-W data sample, $W \to \mu\nu_\mu$ simulation sample and punch through kaons and pions. Points with error bars are for the Pseudo-W data sample, the histogram with the solid (dash) line is for the $W \to \mu\nu_\mu$ simulation (punch through) sample. The punch through sample has been obtained from randomly triggered events with the MBNoBias stripping line where the fraction of muons in these events is assumed to be extremely small (about 0.04%). This fraction is estimated with a simulation of minimum bias events with PYTHIA. Thus a pure sample of kaons or pions can be selected from the MBNoBias strip. In order to avoid low statistics, the p_T requirement on these kaons and pions is $p_T > 5$ GeV/c, rather than $p_T > 20$ GeV/c.

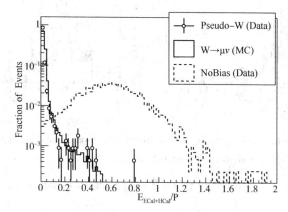

Figure 5.13 Relative energy deposition distributions for the Pseudo-W data sample, $W \to \mu\nu_\mu$ simulation sample and punch through kaons or pions. These three distributions have been normalized to unit area

The relative energy deposition distribution for the Pseudo-W data sample shows an underestimate of the same distribution for the $W \to \mu \nu_\mu$ simulation. This is due to the reason that the p_T spectrum is harder in the Pseudo-W data sample than in the $W \to \mu \nu_\mu$ simulation sample. There is a peak around 0 for both distributions of the Pseudo-W data sample and $W \to \mu \nu_\mu$ simulation sample. The shape of punch through kaons and pions relative energy deposition is assumed to be a Gaussian distribution. This assumption could be verified by applying a Gaussian distribution fit to the relative energy deposition distributions in MBNoBias data with different p_T thresholds on the pions and kaons, such as $p_T > 7$, 9 and 11 GeV/c. Figure. 5.14 shows the fit results. The Gaussian distributions well describe the shapes of these relative energy deposition distributions.

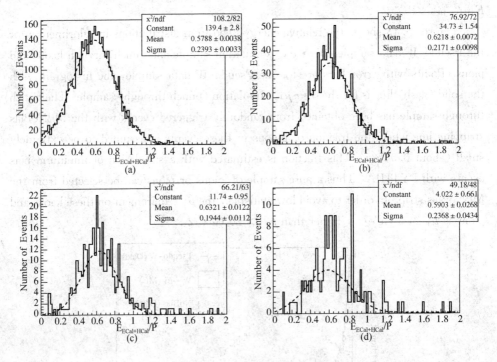

Figure 5.14 (a) ((b), (c), (d)), the relative energy deposition distribution in MBNoBias data with $p_T > 5$ (7, 9, 11) GeV/c is shown as the open histogram, the Gaussian fit is shown as the dash curve

A cut with $E_{ECAL+HCAL}/P < 0.04c$ [1] is applied in order to suppress the punch

---

[1] This cut is optimized to get the maximum signal selection efficiency and purity. The selection efficiency is defined as the fraction of signal events passed this cut. The purity is defined as NS/(NS+NB), where NS is the number of signal events passing the cut, NB is the number of background events passing the cut.

through kaons or pions background. After applying this cut, about 93% of punch through background events are removed while about 92% of signal events are kept.

5.4.4 "Charge" p_T and "neutral" p_T cut in a cone

Muons from the W bosons decays tend to be isolated while muons from the B- or C-hadrons decay (heavy flavor decay) tend to be surrounded by other particles. In order to suppress the muons from the heavy flavor decay, a cone around the muon candidate track is considered. This cone is defined in the (η, ϕ) space. Here η is the pseudo-rapidity, ϕ is the azimuthal angle. The transverse momentum of the vector sum of all other charged tracks in that cone ("charged" p_T^{cone}),

$$p_T^{cone} = \sqrt{\left(\sum_{i \neq \mu} p_x^i\right)^2 + \left(\sum_{i \neq \mu} p_y^i\right)^2} \qquad (5.1)$$

is utilized to study the muon isolation. Here p_x^i (p_y^i) is the momentum in the $x(y)$ direction for charged tracks excluding the muon candidate track, μ, in the cone. As described in the end of section 2.2.2, there are some higher order corrections for the W boson production. One important higher order correction is the process with the W boson accompanied with hadrons. These hadrons are called jets. They are produced by the hadronization of quarks or gluons (see (d), (e), (f) and (g) of Figure. 2.8). In order to not reject $W \to \mu \nu_\mu$ events with jets, the radius of the cone is chosen to be 0.5. Any other charged track is included in the cone if the distance in the (η, ϕ) space between the muon candidate track and this charged track is smaller than 0.5. This requirement is written as follows:

$$\sqrt{(\phi_{track} - \phi_\mu)^2 + (\eta_{track} - \eta_\mu)^2} < 0.5 \qquad (5.2)$$

where the subscript "track" refers to charged tracks excluding the muon candidate track. Figure. 5.15 is a schematic plot of the cone around the muon candidate track. In order to investigate neutral particles (e.g. π^0) around the muon candidate track, the transverse momentum of the vector sum of all neutral deposits in the same cone ("neutral" p_T^{cone}) has been considered. Figure. 5.16(a) shows the "charged" p_T distributions in the cone for the Pseudo-W data sample, $W \to \mu \nu_\mu$ simulation sample and heavy flavor sample. The heavy flavor sample is a data driven sample with a requirement of IP > 0.1 mm on data. Points with error bars are for the Pseudo-W data sample, the histogram with the solid (dash) line is for the $W \to \mu \nu_\mu$ simulation (heavy flavour) sample.

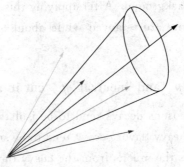

Figure 5.15 A schematic plot of the cone around the muon candidate track. The black arrow is a muon candidate track; the other color arrows are tracks such as kaons or pions

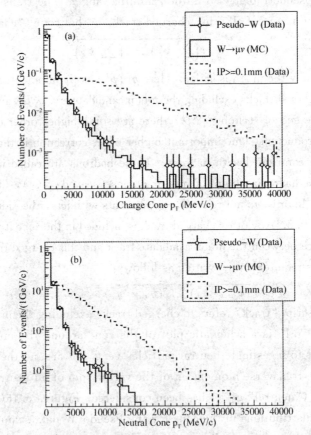

Figure 5.16: (a), "charged" p_T distributions in the cone. (b), "neutral" p_T distributions in the cone. The distributions for the Pseudo-W data sample, $W \to \mu \nu_\mu$ simulation sample and heavy flavor sample are shown on both plots. The distributions in (a) and (b) have been normalized to unit area

Good agreement is achieved between the "charged" p_T^{cone} distributions in the cone for the Pseudo-W data sample and $W \to \mu \nu_\mu$ simulation sample. There are small activities around the muon candidate track in these two samples, while there are much more activities around the muon candidate track from the heavy flavor sample. A cut with "charged" $p_T^{cone} < 2$ GeV/c can remove about 84% of muons from the heavy flavor decay while it can keep about 86% of signal events.

Figure. 5.16(b) shows the "neutral" p_T distributions in the cone for the Pseudo-W data sample, $W \to \mu \nu_\mu$ simulation sample and heavy flavor sample. In order to remove muons from the heavy flavor decays, a cut with "neutral" $p_T^{cone} < 2$ GeV/c is applied on the signal and background samples. With this cut, about 62% of muons from the heavy flavor decay are removed while about 94% of signal events are kept.

5.5 Templates in the fit

As described in section 5.1, $W \to \mu \nu_\mu$ is the signal process in this analysis. The cross-section for this signal process, $\sigma_{W \to \mu \nu_\mu}$, is defined as[89]:

$$\sigma_{W \to \mu \nu_\mu} = \frac{N_{real}^W}{\int L} \quad (5.3)$$

Here N_{real}^W is the real number of signal events in the data sample with an integrated luminosity of $\int L$. However in an experiment the number of observed candidates for the signal process (N_{obs}^W) is different from the real number of events for the signal process. This is due to the following reasons.

First, due to detector inefficiencies, the real number of signal events is underestimated. There are three sources that cause detector inefficiencies. They are the experiment trigger procedure, the imperfect performance of track reconstruction and the particle identification. In order to get the real number of signal events, a correction factor, $1/\varepsilon_{detector}^\mu$, is applied. $\varepsilon_{detector}^\mu$ is the detector efficiency. It is defined as a product of: the track reconstruction efficiency (ε_{track}^μ), describing the probability of reconstructing long tracks with hits in detectors; the identification efficiency (ε_{id}^μ), describing the probability of identifying tracks as muons and the trigger efficiency (ε_{trig}^μ), describing the probability of triggering on such offline signal events. The ways to determine ε_{track}^μ, ε_{id}^μ and ε_{trig}^μ are presented in section 5.7.

Second, due to the application of candidate selection cuts, the real number of signal events is also underestimated. In order to correct this inefficiency, a factor, $1/\varepsilon_{\text{selection}}^{\mu}$ is the selection efficiency. It is the number of signal events within the kinematic phase-space defined by the selection cuts divided by the number of signal events in all phase-space and is written as $\varepsilon_{\text{selection}}^{\mu} = N_{\text{kinematic space}}^{W} / N_{\text{all space}}^{W}$.

Third, the number of signal events is counted in a reduced muon p_T range between 20 GeV/c and 70 GeV/c. In order to calculate the cross-section of the signal process in a muon p_T range with $p_T > 20$ GeV/c, a correction factor, $1/\varepsilon_{\text{acceptance}}^{\mu}$, is applied. Here $\varepsilon_{\text{acceptance}}^{\mu}$ is the acceptance efficiency and is defined as the number of signal events with $20 < p_T < 70$ GeV/c divided by the number of signal events with $p_T > 20$ GeV/c.

Fourth, due to contaminations of signal events with background events, the number of observed candidates for the signal process, N_{obs}^{W}, is different from the real number of signal events. Thus the number of background events (N_{bkg}^{W}) must be subtracted from the number of observed candidates for the signal process.

The final expression for the $\sigma_{W \to \mu\nu_\mu}$ then is as follows:

$$\sigma_{W \to \mu\nu_\mu} = \frac{N_{\text{obs}}^{W} - N_{\text{bkg}}^{W}}{\int L \cdot \varepsilon_{\text{detector}}^{\mu} \cdot \varepsilon_{\text{selection}}^{\mu} \cdot \varepsilon_{\text{acceptance}}^{\mu}} + \sigma_{\text{FSR}}^{W} \qquad (5.4)$$

where the last term, σ_{FSR}^{W}, is a QED Final State Radiation (FSR) correction to the W cross-section. This term corrects for photon emission from the W final states. Photon emission is present in the W cross-section measurement in data, while it is not present in the theoretical prediction. In order to make a consistent comparison with the theoretical prediction, the measured cross-section in data is corrected to Born level in QED.

In order to estimate how many signal and background events in the data sample, the shapes of p_T distributions in the signal and background samples are fitted to the shape of the p_T distribution in the data sample. The fit is based on an extended likelihood method❶[100] and is performed with a TFractionFitter Root package[101]. The signal and background templates in the fit are described in the following sections.

5.5.1 Heavy flavor sample

As there are very low statistics in the signal region ($p_T > 20$ GeV/c) for $b\bar{b} \to X\mu$ and $c\bar{c} \to X\mu$ simulation samples, a data driven sample has been produced

---

❶ This method takes into account both data and Monte Carlo statistical uncertainties. These uncertainties are treated as Poisson errors.

with the following method for the semi-leptonic decays of heavy flavor hadrons containing $b(\bar{b})$ or $c(\bar{c})$ quarks. Figure. 5.12 shows that with a cut of IP $>$ 100 μm there are very few signal $W \to \mu \nu_\mu$ events while there are lots of $b\bar{b} \to X\mu$ and $c\bar{c} \to X\mu$ events. A heavy flavor data driven sample has been derived by applying the cut with IP $>$ 100 μm on the data sample❶.

The shape of the p_T distribution in the heavy flavor sample is taken from this data driven sample. In order to get the normalization of the heavy flavor sample, the IP distributions of signal and background templates are fitted to the IP distribution of the data sample with the TFractionFitter Root package. As the Pseudo-W data sample gives a consistent description for the $W \to \mu \nu_\mu$ simulation sample, the signal sample in the fit is taken from the Pseudo-W data sample. The heavy flavor sample in the fit is taken from the combination of $b\bar{b} \to X\mu$ and $c\bar{c} \to X\mu$ simulation samples. All candidate selection cuts except the IP cut have been applied on the data sample, Pseudo-W data sample and the combination of $b\bar{b} \to X\mu$ and $c\bar{c} \to X\mu$ simulation samples. In order to get enough statistics, there is no transverse momentum cut applied on muon tracks, as their impact parameters do not depend on their transverse momenta. Figure. 5.17 shows the fit result. The χ^2/ndf of the fit is 0.33. With this method the fraction of heavy flavor sample is $(0.6 \pm 0.2)\%$.

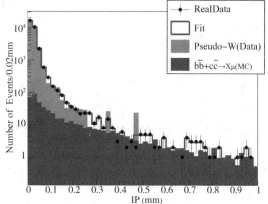

Figure 5.17 Fitting the impact parameter distributions in the Pseudo-W data sample and $b\bar{b}+c\bar{c} \to X\mu$ sample to the impact parameter distribution in the data sample with TFractionFitter. Points with error bars are for the data sample. The grey (black) histogram is for the Pseudo-W data ($b\bar{b}+c\bar{c} \to X\mu$ simulation) sample. The open histogram is the fit result

---

❶ The data sample is a sample in which events have passed the W2Mu stripping line in section 4.5. This data sample is a mixture of signal and background samples.

5.5.2 Punch through sample

Kaons or pions can punch through the ECAL and HCAL into muon chambers and thus they can fake muons. As there are very low statistics in the signal region ($p_T > 20$ GeV/c) for the punch through sample which is derived from randomly triggered events with the MBNoBias stripping line, a data driven sample is produced with the following method. As described in section 5.4.3, muons typically deposit a small amount of energy in calorimeters while hadrons deposit much more energy than muons. Figure. 5.13 shows that when $E_{ECAL+HCAL}/P > 0.3c$, there are very few muon candidates, while there are lots of pions and kaons, thus a punch through data driven sample can be derived by applying a requirement with $E_{ECAL+HCAL}/P > 0.3c$ on the data sample. Figure. 5.18 shows the relative energy deposition distribution of muon candidates in the data sample. These muon candidates have fired the trigger lines described in section 4.4.5.

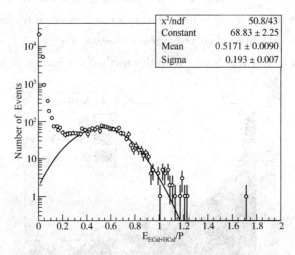

Figure 5.18 Relative energy deposition distribution of muon candidates in the data sample. The black curve is a Gaussian fit

The shape of the p_T distribution in the punch through sample is taken from the data driven sample. As described in section 5.4.3, the shape of the relative energy distribution for pions and kaons could be described by a Gaussian distribution. Thus in order to estimate how many punch through background events are expected in the signal region ($E_{ECAL+HCAL}/P < 0.04c$), a Gaussian fit is applied to the relative energy deposition distribution in the data sample. The fit range is from 0.3c to 2c. This fit is shown as the black curve in Figure. 5.18. This curve is extrapolated to the signal region. The number of expected punch through background events in the signal region then is obtained by integrating the Gaussian fit with the relative energy deposition from 0c to 0.04c. The normalization of the punch through sample is estimated to be the number of expected punch through background events divided by the number of observed W candidates in the data sample. This normalisation is $(0.19\pm0.08)\%_0$.

5.5.3 Decay in flight sample

Pions and kaons can decay in flight into muons and neutrinos somewhere between the proton-proton interaction point and the ECAL (see Figure. 3.3). As these muons deposit a small amount of energy in calorimeters, they are dominant backgrounds in the signal region with $E_{\text{ECAL+HCAL}}/P < 0.04c$.

In order to avoid low statistics, a decay in flight data driven sample is produced to study this background. Tracks in this sample are selected from the W2MuNoPIDs stripping line. In order to make this sample unbiased to any given process, it is required that each track in this sample does not fire any trigger lines (TIS, trigger independent of signal). As no identification information is associated with tracks in this sample and the transverse momenta of these tracks are greater than 15 GeV/c, this sample is deemed as a similar minimum bias sample with high p_T particles in the final state. The p_T spectrum of muons from the pion or kaon decay is produced by weighting each TIS track with its probability to decay into a muon. A minimum bias sample with randomly triggered events selected from the MBNoBias stripping line is utilized to determine the probability for pions and kaons to fake muons, $\text{Prob}_{\text{mis-ID}}$. This probability is defined as the fraction of tracks in the minimum bias sample identified as muons. Its distribution is fitted with the following function[72]:

$$\text{Prob}_{\text{mis-ID}} = (1 - e^{-\frac{p_0}{P}}) + (p_1 \cdot P + p_2) \quad (5.5)$$

where p_0, p_1 and p_2 are parameters in the fit, P is the hadron momentum. The first term is an exponential function, which describes the probability of pions and kaons decaying in flight into muons; the second term is a linear term, which describes the probability of kaons or pions punching through the calorimeters into the muon chambers. Only the first term is used to weight the TIS tracks.

Figure. 5.19 shows the misidentification probability distribution as a function of hadron momenta in the whole η range for positive and negative hadrons. The fit function is shown as the black curve in the plot. Empty bins are ignored in the fit, thus the black curve on the empty bins is an extrapolation of the fit. The misidentification probability distributions in each η bin are shown in Figure. 5.20. As the punch through component contributes at higher momentum regions and there are not enough statistics at these regions for each η bin, the misidentification probability due to this component is constrained in the fit for each η bin. For positive (negative) hadrons the parameters p_1 and p_2 in each η bin are limited in the ranges of $7.4 \times 10^{-5} < p_1 < 11.2 \times 10^{-5}$ and $-0.006 < p_1 < -0.004$ ($4.2 \times 10^{-5} < p_1 < 8.2 \times 10^{-5}$ and $-0.004 < p_1 < -0.002$). These ranges are taken from the fit results in the whole η range.

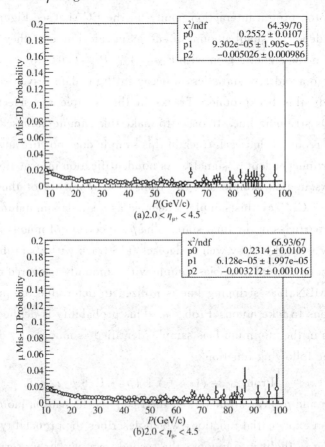

Figure 5.19 The probability of pions and kaons decaying into muons as a function of their momenta in the whole η range. Points with error bars show the probability. The black line is a fit to the probability using the function described in Eq. (5.5). The upper plot is for the positive muons, the lower plot is for the negative muons

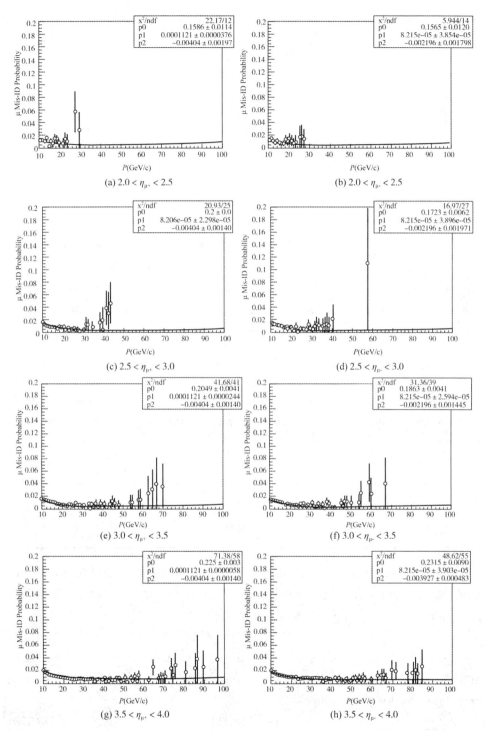

Figure 5.20 Probabilities of pions and kaons decaying into muons as a function of their momenta in each η bin. The left (right) column is for positive (negative) hadrons

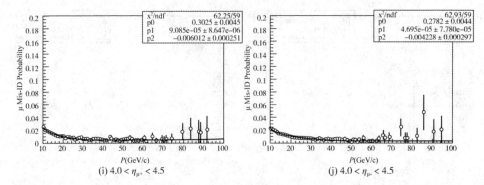

Figure 5.20(continue) Probabilities of pions and kaons decaying into muons as a function of their momenta in each η bin. The left (right) column is for positive (negative) hadrons

Table 5.1 shows p_0 values in five η bins for positive and negative hadrons. These p_0 values are inconsistent between positive and negative hadrons in the second, third and fifth η bins. They are also inconsistent between η bins for positive and negative hadrons. This inconsistency between η bins could be explained theoretically as follows. The mean decay length of pions is[72]

$$c\tau_\pi^{Lab} \approx \frac{c\tau_\pi^{Rest}}{m_\pi} \cdot P_\pi \qquad (5.6)$$

Table 5.1 p_0 values in five η bins. p_0^+ is for the positive hadrons. p_0^- is for negative hadrons

η	p_0^+	p_0^-
2.0—2.5	0.16±0.01	0.16±0.01
2.5—3.0	0.200±0.007	0.172±0.006
3.0—3.5	0.205±0.004	0.186±0.004
3.5—4.0	0.225±0.003	0.232±0.009
4.0—4.5	0.303±0.005	0.278±0.004

where c is the speed of light in vacuum, τ_π^{Lab} (τ_π^{Rest}) is the π mean life time in the Lab (Rest) frame, m_π is the mass of π, P_π is the momentum of π. Thus the probability of decay in fight in a region Δz for pions is

$$1 - e^{-\Delta z / c\tau_\pi^{Lab}} = 1 - e^{-\Delta z / 55.8 P_\pi} \qquad (5.7)$$

In the LHCb detector, the maximum value of Δz is 15 m, thus the maximum p_0 value for pions is 0.269. In the same way, we can get the decay in fight probability for kaons as follows:

$$1 - e^{-\Delta z / 7.5 P_K} \qquad (5.8)$$

where P_K is the momentum of kaons. The maximum p_0 value for kaons is 1.994.

As the decay in flight sample is a mixture of pions and kaons, the maximum value of p_0 should be between 0.269 and 1.994. p_0 values in Table 5.1 prove this statement. The pseudo-rapidity, η, is a function of the polar angle between the hadron momentum and the beam axis, θ : $\eta = -\ln\left[\tan\frac{\theta}{2}\right]$. For different η bins, θ values are different and thus Δz values are different. As a result, the decay in flight probability in each η bin is different.

In each η bin, TIS tracks are reweighted with the $\text{Prob}_{\text{mis-ID}}$ assigned to that η bin. Five templates are utilized to describe the decay in flight sample. Each template corresponds to the decay in flight sample in one η bin. The normalization for each template is free to vary in the p_T spectrum fit.

5.5.4 $W \to \mu\nu_\mu$ simulation sample

The LHCb simulation software of Gauss, Boole, Brunel and Moore described in section 4.6 are utilized to produce a $W \to \mu\nu_\mu$ simulation sample. As described in section 4.6.1, the default generator in Gauss is PYTHIA 6.4. Thus events in the $W \to \mu\nu_\mu$ simulation sample contain leading-order (LO) information. Another set of $W \to \mu\nu_\mu$ events have been generated by the POWHEG generator at next-leading-order (NLO). Only events generated at LO by PYTHIA are fully simulated through the LHCb software. In order to get fully simulated events at NLO, the p_T spectrum generated by PYTHIA at reconstruction level is reweighted with a K-factor distribution. The K-factor is defined as the ratio between the cross-section calculated up to LO for a process and the cross-section calculated up to NLO with the same process[102] and it is written as $K = \dfrac{\sigma_{NLO}}{\sigma_{LO}}$.

Figure. 5.21 shows muon p_T spectra for the $W \to \mu\nu_\mu$ process at truth level in the POWHEG (solid line) and PYTHIA (dash line) samples. The muon spectrum is split into five η bin. In each η bin, the left plot is for the positive muon spectrum while the right plot is for the negative muon spectrum. The K-factor then is calculated as the ratio between the number of events at truth level in the POWHEG and PYTHIA samples. Figure. 5.22 shows the K-factor distribution. This distribution is split into five η bin. In each η bin, the left plot is for the positive muon while the right plot is for the negative muon. The shaded band is the uncertainty associated to the K-factor. As there are low statistics in the high η bin and in the high p_T regions, the K-factors in these regions fluctuate a lot and the uncertainties on the K-factors are large.

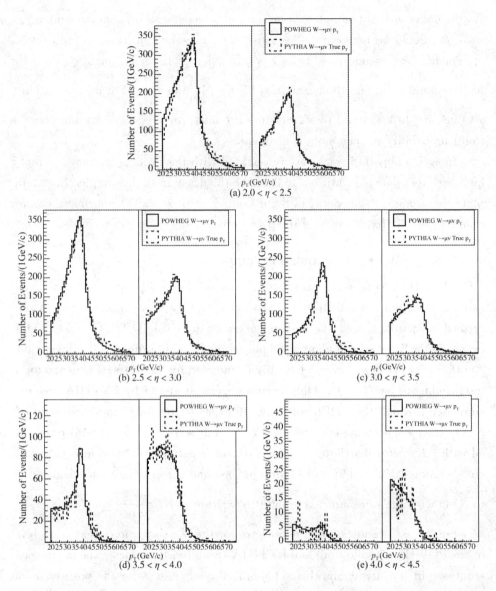

Figure 5.21 Muon p_T distributions at truth level for the $W \to \mu\nu_\mu$ process in the POWHEG (solid line) and PYTHIA (dash line) samples. In each η bin, the left (right) is for positive (negative) muons

Figure. 5.23 shows the muon p_T spectrum generated by PYTHIA at reconstruction level for the $W \to \mu\nu_\mu$ process. This muon p_T spectrum is reweighted with the K-factor distribution bin by bin in order to get a spectrum at reconstruction level with NLO information. The histogram with the solid (dash) line is the muon p_T spectrum before (after) reweighting. The p_T spectrum at reconstruction level is also split into five η bin, the left plot is for the positive muon while the right plot is for the negative muon. Only the reweighted muon p_T spectrum at reconstruction level is used as a template in the p_T spectrum fit and its normalization is free to vary in that fit.

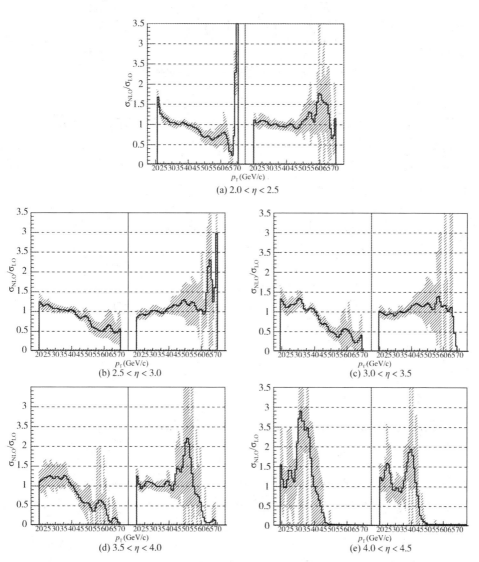

Figure 5.22 The histograms with solid lines are K-factor distributions in five η bins for the $W \to \mu \nu_\mu$ process. In each η bin, the left (right) is for positive (negative) muons. The uncertainties on the K-factors are shown as shaded bands

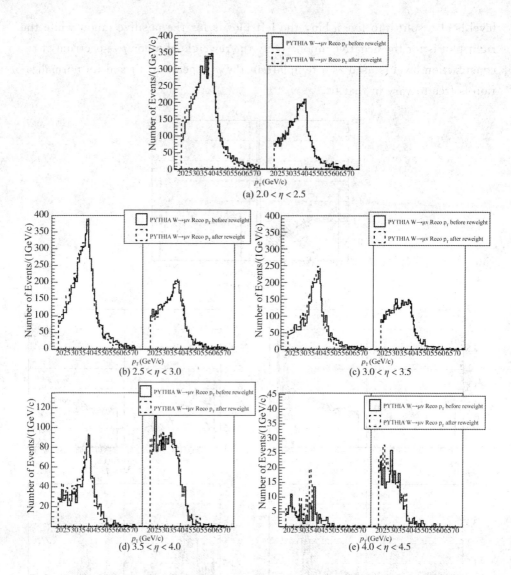

Figure 5.23 Muon p_T distributions generated at reconstruction level before (solid line) and after (dash line) reweighting for the $W \to \mu \nu_\mu$ process. In each η bin, the left (right) is for positive (negative) muons

5.5.5 $Z \to \mu\mu$ simulation sample

The LHCb simulation software is also utilized to produce a $Z \to \mu\mu$ simulation sample. In this sample, there is only one high p_T muon in the LHCb acceptance in one event. $Z \to \mu\mu$ events generated by PYTHIA 6.4 contain LO information. Another set of $Z \to \mu\mu$ events are generated at NLO with the POWHEG generator. The same procedure in section 5.5.4 is utilized here to get a p_T spectrum at reconstruction level with NLO information. Figure. 5.24 shows the K-factor distribution. The histogram with the solid (dash) line in Figure. 5.25 is the muon p_T spectrum before (after) reweighting.

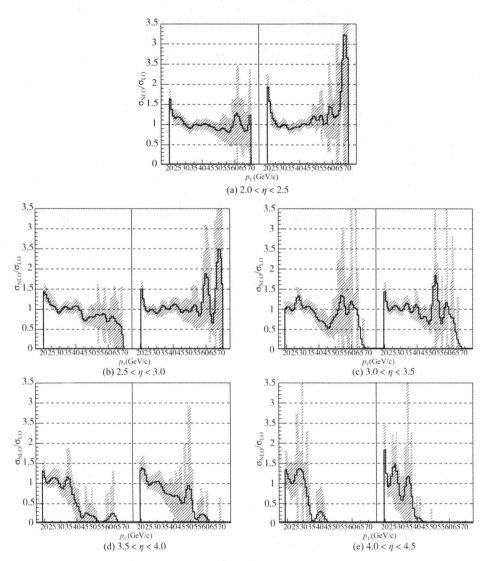

Figure 5.24 The histograms with solid lines are K-factor distributions in five η bins for the $Z \to \mu\mu$ process. In each η bin, the left (right) is for positive (negative) muons. The uncertainties on the K-factors are shown as shaded bands

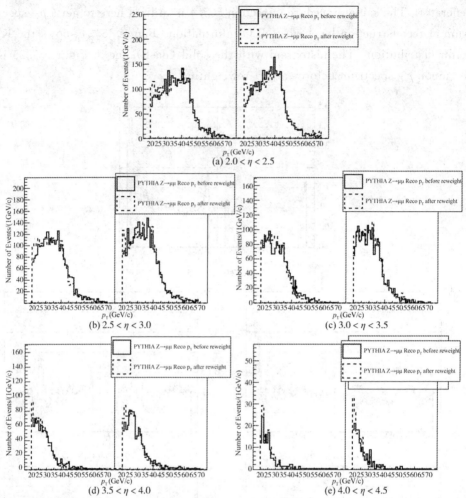

Figure 5.25 Muon p_T distributions generated at reconstruction level before (solid line) and after (dash line) reweighting for the $Z \to \mu\mu$ process. In each η bin, the left (right) is for positive (negative) muons

The $Z \to \mu\mu$ simulation sample normalization is determined as the number of reconstructed $Z \to \mu\mu$ events which pass through candidate selection cuts and are expected in the data sample, $N^{Z \to \mu\mu}_{\text{expected in data}}$, divided by the number of observed W candidates in the data sample. $N^{Z \to \mu\mu}_{\text{expected in data}}$ is determined as follows:

$$N^{Z \to \mu\mu}_{\text{expected in data}} = L \cdot \sigma_{Z \to \mu\mu} \cdot \varepsilon^{Z \to \mu\mu}_{\text{genarator}} \cdot \varepsilon_{\text{GEC}} \cdot N^{Z \to \mu\mu}_{\text{expected in MC}} / N^{Z \to \mu\mu}_{\text{expected in LHCb}} \quad (5.9)$$

Here L is the integrated luminosity of the data sample. $\sigma_{Z \to \mu\mu}$ is the $Z \to \mu\mu$ process cross-section predicted at NNLO with the MSTW08 PDF set[103]. ε_{GEC} is the effciency due to the global event cuts. It is defined as the fraction of events in the data sample which pass through the global event cuts. $N^{Z \to \mu\mu}_{\text{expected in MC}}$ is the number of events observed in the $Z \to \mu\mu$ simulation sample in which candidate selection cuts and trigger requirements are applied. $\varepsilon_{\text{genarator}}^{Z \to \mu\mu}$ is a generator acceptance factor for the $Z \to \mu\mu$ simulation. It is defined as the number of events accepted in the LHCb acceptance, $N_{\text{expected in LHCb}}^{Z \to \mu\mu}$, divided by the number of events generated in 4π space by PYTHIA, $N_{\text{expected in } 4\pi}^{Z \to \mu\mu}$ and is written as

$$\varepsilon_{\text{genarator}}^{Z \to \mu\mu} = \frac{N_{\text{expected in LHCb}}^{Z \to \mu\mu}}{N_{\text{expected in } 4\pi}^{Z \to \mu\mu}} \quad (5.10)$$

Plugging Eq. (5.10) into Eq. (5.9), we can rewrite $N_{\text{expected in data}}^{Z \to \mu\mu}$ as follows:

$$N_{\text{expected in data}}^{Z \to \mu\mu} = L \cdot \sigma_{Z \to \mu\mu} \cdot N_{\text{expected in MC}}^{Z \to \mu\mu} / N_{\text{expected in } 4\pi}^{Z \to \mu\mu} \cdot \varepsilon_{\text{GEC}} \quad (5.11)$$

Inserting $N_{\text{expected in data}}^{Z \to \mu\mu}$ into the expression of the $Z \to \mu\mu$ simulation sample normalization, we get this normalization as $(8.0 \pm 0.5 \pm 0.9)\%$. The first uncertainty is the statistical uncertainty. The second is systematic uncertainty and is taken from the difference between the normalizations estimated from simulation and data[51].

5.5.6 $W \to \tau\nu_\tau$ simulation sample

The $W \to \tau\nu_\tau$ simulation sample is also produced by the LHCb software. In this sample, τ decays into a muon and neutrinos. The normalisation of the $W \to \tau\nu_\tau$ simulation sample is determined in the same way as the $Z \to \mu\mu$ simulation sample. It is $(2.0 \pm 0.2)\%$.

5.5.7 $Z \to \tau\tau$ simulation sample

The $Z \to \tau\tau$ simulation sample is also produced by the LHCb software. In this sample, only one τ decays into a muon and neutrinos in the LHCb acceptance. The normalization of the $Z \to \tau\tau$ simulation sample is determined in the same way as the $Z \to \mu\mu$ simulation sample. It is $(0.39 \pm 0.02)\%$.

The generator acceptance factors and cross-sections predicted at NNLO with the MSTW08 PDF set for the $Z \to \mu\mu$, $W \to \tau\nu_\tau$ and $Z \to \tau\tau$ processes are listed in Table 5.2.

Table 5.2 Generator acceptance factors and cross-sections for the $Z \to \mu\mu$, $W \to \tau\nu_\tau$ and $Z \to \tau\tau$ processes

Simulation sample	Cross-section (mb)	$\varepsilon_{\text{generator}}$
$Z \to \mu\mu$	$(9.6 \pm 0.3) \times 10^{-7}$	$(37.0 \pm 0.9)\%$
$W \to \tau\nu_\tau$	$(10.5 \pm 0.3) \times 10^{-6}$	$(24.6 \pm 0.7)\%$
$Z \to \tau\tau$	$(9.6 \pm 0.3) \times 10^{-7}$	$(36.6 \pm 0.4)\%$

5.6　Fit results

A TFractionFitter fit to the p_T spectrum has been carried out to determine the purity of the signal $W \to \mu \nu_\mu$ sample. The way to perform the fit is described as follows.

The data sample is described with one histogram (h_1). The signal sample is described with 10 histograms (h_2 to h_{11}). The decay in flight background sample is described with 5 histograms (h_{12} to h_{16}). Other background samples ($Z \to \mu\mu$, heavy flavor, punch through, $W \to \tau \nu_\tau$ and $Z \to \tau\tau$) are described with 5 histograms (h_{17} to h_{21}), one for each background sample. For each of these 21 histograms, the x-axis represents the muon's p_T, the y-axis represents the number of events with the muon's p_T in a given p_T range. The x range of the histogram starts from 15 GeV/c and ends at 615 GeV/c, and the number of bins in the histogram is 600. In the data sample, the p_T of the positive (negative) muon with $2.0 < \eta^\mu < 2.5$ is added with 0 (60) GeV/c and then fills h_1 in the x range 20-70 (80-130) GeV/c; the p_T of the positive (negative) muon with $2.5 < \eta^\mu < 3.0$ is added with 120 (180) GeV/c and then fills h_1 in the x range 140-190 (200-250) GeV/c; the p_T of the positive (negative) muon with $3.0 < \eta^\mu < 3.5$ is added with 240 (300) GeV/c and then fills h_1 in the x range 260-310 (320-370) GeV/c; the p_T of the positive (negative) muon with $3.5 < \eta^\mu < 4.0$ is added with 360 (420) GeV/c and then fills h_1 in the x range 380-430 (440-490) GeV/c; the p_T of the positive (negative) muon with $4.0 < \eta^\mu < 4.5$ is added with 480 (540) GeV/c and then fills h_1 in the x range 500-550 (560-610) GeV/c.

For the $W \to \mu\nu$ signal sample, the p_T of the positive (negative) muon with $2.0 < \eta^\mu < 2.5$ is added with 0 (60) GeV/c and then fills h_2 (h_3) in the x range 20-70 (80-130) GeV/c; the p_T of the positive (negative) muon with $2.5 < \eta^\mu < 3.0$ is added with 120 (180) GeV/c and then fills h_4 (h_5) in the x range 140-190 (200-250) GeV/c; the p_T of the positive (negative) muon with $3.0 < \eta^\mu < 3.5$ is added with 240 (300) GeV/c and then fills h_6 (h_7) in the x range 260-310 (320-370) GeV/c; the p_T of the positive (negative) muon with $3.5 < \eta^\mu < 4.0$ is added with 360 (420) GeV/c and then fills h_8 (h_9) in the x range 380-430 (440-490) GeV/c; the p_T of the positive (negative) muon with $4.0 < \eta^\mu < 4.5$ is added with 480 (540) GeV/c and then fills h_{10} (h_{11}) in the x range 500-550 (560-610) GeV/c.

For the decay in flight sample, the p_T of the positive (negative) muon with

$2.0 < \eta^\mu < 2.5$ is added with 0 (60) GeV/c and then fills h_{12} in the x range 20-70 (80-130) GeV/c; the p_T of the positive (negative) muon with $2.5 < \eta^\mu < 3.0$ is added with 120 (180) GeV/c and then fills h_{13} in the x range 140-190 (200-250) GeV/c; the p_T of the positive (negative) muon with $3.0 < \eta^\mu < 3.5$ is added with 240 (300) GeV/c and then fills h_{14} in the x range 260-310 (320-370) GeV/c; the p_T of the positive (negative) muon with $3.5 < \eta^\mu < 4.0$ is added with 360 (420) GeV/c and then fills h_{15} in the x range 380-430 (440-490) GeV/c; the p_T of the positive muon with $4.0 < \eta^\mu < 4.5$ is added with 480 (540) GeV/c and then fills h_{16} in the x range 500-550 (560-610) GeV/c.

The way to fill the histograms h_{17}-h_{21} is the same as the way to fill h_1. The p_T spectrum in h_1 is fitted as the sum of the p_T spectra in histograms h_2-h_{21}. The fractions of h_2-h_{11} in h_1 are free to vary in the fit, each fraction takes one free parameter. The fractions of h_{12}-h_{16} in h_1 are also free to vary in the fit, each fraction takes one free parameter. The fractions of h_{17}-h_{21} in h_1 are constrained in the fit.

As a summary, this fit utilizes 15 free parameters to describe the signal and decay in flight background templates. The signal template is described with 10 parameters in which five parameters are for the positive muon p_T spectra in five η bins while the other five parameters are for the negative muon p_T spectra in five η bins. The decay in flight template is described with 5 free parameters, each for one η bin. Since the decay in flight sample is produced by reweighting the TIS track in the W2MuNoPID line with the probability of kaons and pions decaying in flight into muons, the ratio between the positive muon and negative muon in the decay in flight sample should be the same as the ratio between the TIS tracks with the positive charge and negative charge. Other background template fractions in the data sample are constrained with their normalizations. The methods to determine background normalizations are described in section 5.5. Figure. 5.26 shows the template fit result. This fit result is split into five subplots. Each subplot is for one η bin. Points with error bars are for the data sample. The fit result is shown as the histogram with the grey line. Signal and background contributions are indicated by the key.

The fractions of h_2-h_{11} (for the signal sample) and h_{12}-h_{16} (for the decay in flight template) are returned by the fit. The overall fraction of the signal template is determined by adding the fractions of h_2-h_{11} together. The overall fraction of the decay in flight template is determined by adding the fractions of h_{12}-h_{16} together. The overall fractions of signal and decay in flight templates are tabulated in Table 5.3. The purity of muons from the W bosons is about 79%. Table

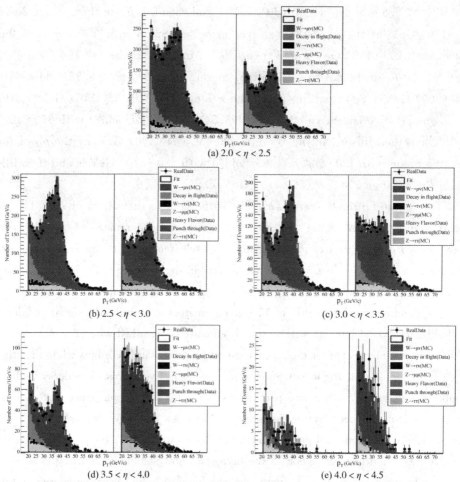

Figure 5.26 p_T distributions of signal and background templates are fitted to the p_T distribution of the data sample in five η bins. In each subplot, the left (right) histogram is for the positive (negative) muon p_T spectrum.

The signal and background contributions are indicated by the key. In the key, "Data" refers to a data-driven sample, "MC" refers to a simulation sample

5.4 shows the number of observed candidates for the $W^+ \to \mu^+ \nu_\mu$ process, $N_{obs}^{W^+ \to \mu^+ \nu_\mu}$, and for the $W^- \to \mu^- \bar{\nu}_\mu$ process, $N_{obs}^{W^- \to \mu^- \bar{\nu}_\mu}$, in each η bin for the data sample. The uncertainty quoted is the statistical uncertainty and it is taken as a square root of the number of observed candidates in each η bin for positve and negative muons. Table 5.5 shows the number of events for the $W^+ \to \mu^+ \nu_\mu$ background, $N_{bkg}^{W^+ \to \mu^+ \nu_\mu}$, and for the $W^- \to \mu^- \bar{\nu}_\mu$ background, $N_{bkg}^{W^- \to \mu^- \bar{\nu}_\mu}$, in five η bins. The uncertainty quoted is the statistical uncertainty of the decay in flight sample returned by the fit. As the decay in flight sample is described with five histograms h_{12}-h_{16}, and these five histograms are utilized in one fit, in order to make the fit successful, the increase of the fraction for one histogram will result the increase of fractions for the other four histograms. As a result, we assume that the uncertainties on the number of background events returned by the fit are fully correlated between η bins.

Table 5.3 Fractions of the signal and decay in flight background templates in the data samples

Sample	Fraction	Sample	Fraction
$W^+ \to \mu^+ \nu_\mu$	$(44.5 \pm 1.2)\%$	Decay in flight	$(9.6 \pm 0.8)\%$
$W^- \to \mu^- \bar{\nu}_\mu$	$(34.8 \pm 1.1)\%$	χ^2/ndf	1.03

Table 5.4 Numbers of observed $W^+ \to \mu^+ \nu_\mu$ and $W^- \to \mu^- \bar{\nu}_\mu$ candidates in each η bin for the data sample

η	$N_{obs}^{W^+ \to \mu^+ \nu_\mu}$	$N_{obs}^{W^- \to \mu^- \bar{\nu}_\mu}$	η	$N_{obs}^{W^+ \to \mu^+ \nu_\mu}$	$N_{obs}^{W^- \to \mu^- \bar{\nu}_\mu}$
2.0 − 2.5	5400±73	3354±58	3.5 − 4.0	1156±34	1688±41
2.5 − 3.0	5169±72	3658±60	4.0 − 4.5	122±11	254±16
3.0 − 3.5	3183±56	2907±54			

Table 5.5 Numbers of $W^+ \to \mu^+ \nu_\mu$ and $W^- \to \mu^- \bar{\nu}_\mu$ background events in five η bins

η	$N_{bkg}^{W^+ \to \mu^+ \nu_\mu}$	$N_{bkg}^{W^- \to \mu^- \bar{\nu}_\mu}$	η	$N_{bkg}^{W^+ \to \mu^+ \nu_\mu}$	$N_{bkg}^{W^- \to \mu^- \bar{\nu}_\mu}$
2.0 − 2.5	1072±35	948±29	3.5 − 4.0	276±23	217±12
2.5 − 3.0	939±33	741±21	4.0 − 4.5	46±11	38±5
3.0 − 3.5	721±34	545±20			

5.7 Muon track detector efficiency

As described at the beginning of section 5.5, in order to get the real number of signal events, the correction factor, $\varepsilon_{detector}^\mu$, is applied. $\varepsilon_{detector}^\mu$ is the detector efficiency. It is a product of: the track reconstruction efficiency (ε_{track}^μ), the identification efficiency (ε_{id}^μ), and the trigger efficiency (ε_{trig}^μ). The determinations of ε_{track}^μ, ε_{id}^μ and ε_{trig}^μ are shown in the following sections.

5.7.1 Muon track identification efficiency

The muon track identification efficiency, ε_{id}^μ, is measured with a tag-and-probe method[89] using a $Z \to \mu\mu$ data sample. This sample is produced with the Z0MuMuNoPIDs stripping line. In the tag-and-probe method, tag tracks are defined as tracks which satisfy track pre-selection requirements as well as the muon identification requirement, IsMuon=1. It is also required that tag tracks should fire the trigger lines and their transverse momenta should be greater than 20 GeV/c. Probe tracks are defined as tracks which satisfy track pre-selection requirements and give an invariant mass of the dimuon pair in the range of 81 GeV/$c^2 < m_{\mu\mu} < 101$ GeV/c^2 when combined with opposite charges of tag tracks. The momenta of probe tracks should also be greater than 20 GeV/c. The invariant mass distribution of dimuon pairs with tagging on μ^- (μ^+) and probing on μ^+ (μ^-) is shown in Figure. 5.27(a) (Figure. 5.27(b)).

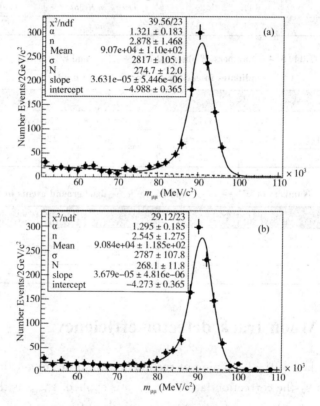

Figure 5.27 (a), the invariant mass distribution of dimuon pairs with tagging on μ^- and probing on μ^+ in the $Z \to \mu\mu$ data sample. (b) the invariant mass distribution of dimuon pairs with tagging on μ^+ and probing on μ^- in the $Z \to \mu\mu$ data sample. The data is shown as points with error bars. The solid (dashed) curve is the crystal ball function (exponential function) fit

This mass distribution is fitted with a crystal ball function plus an exponential function. The crystal ball function is given by[104]:

$$f(x;\alpha,n,\bar{x},\sigma)=N\cdot\begin{cases}\exp\left(-\dfrac{(x-\bar{x})^2}{2\sigma^2}\right) & \text{for } \dfrac{(x-\bar{x})}{2\sigma}>-\alpha \\ A\cdot\left(B-\dfrac{x-\bar{x}}{\sigma}\right)^{-n} & \text{for } \dfrac{(x-\bar{x})}{2\sigma}\leqslant -\alpha\end{cases} \quad (5.12)$$

where $A=\left(\dfrac{n}{|\alpha|}\right)^n\cdot\exp\left(-\dfrac{|\alpha|^2}{2}\right)$, $B=\dfrac{n}{|\alpha|}-|\alpha|$, N is a normalization factor, $\alpha, n, \bar{x}$ and σ are parameters which are fitted to the $Z\to\mu\mu$ data sample. The exponential function is given by:

$$f(x;\alpha,\beta)=\exp(-(\alpha\cdot x+\beta)) \quad (5.13)$$

where α, β are parameters which are fitted to the $Z\to\mu\mu$ data sample. The fits are shown in Figure. 5.27(a) and Figure. 5.27(b). The dominant component of the invariant mass distribution is the Z peak and it is described by the crystal ball function (solid curve). The remnant component represents the background contribution and it is described by the exponential function (dash curve). As the invariant mass of dimuon pairs is in the range 81 GeV/c^2 < $m_{\mu\mu}$ < 101 GeV/c^2, most of background events are removed.

The muon identification efficiency is defined as the fraction of probe tracks which are identified as muons with the IsMuon requirement. The identification efficiencies for positive muons, $\varepsilon_{id}^{\mu^+}$, and negative muons, $\varepsilon_{id}^{\mu^-}$, in five η bins are tabulated in Table 5.6 and they are shown as circle dots with error bars in Figure. 5.28. The application of the identification procedure is considered as a binomial process. Thus the uncertainties quoted in Table 5.6 are binomial errors. As $\varepsilon_{id}^{\mu^+}$ is consistent with $\varepsilon_{id}^{\mu^-}$ in each η bin, a charge unbiased efficiency, ε_{id}^{μ}, is calculated and it is tabulated in the fourth column of Table 5.6. Only ε_{id}^{μ} is utilized in the W cross-section calculation. In order to make a comparison between data and simulation, another set of muon identification efficiencies (square dots with error bars) are calculated with a $Z\to\mu\mu$ simulation sample. General agreement is achieved between efficiencies calculated from data and simulation.

Table 5.6 Identification efficiencies in five η bins. In the third η bin, ε_{id}^{μ} is calculated by inverting the probability in the binomial process with Bayes' Theore[105]. Other uncertainties are binomial errors

η	$\varepsilon_{id}^{\mu^-}$	$\varepsilon_{id}^{\mu^+}$	ε_{id}^{μ}
2.0—2.5	(98.7 ± 0.7)%	(99.1 ± 0.5)%	(98.9 ± 0.4)%
2.5—3.0	(99.0 ± 0.6)%	(98.4 ± 0.7)%	(98.7 ± 0.5)%

continue table

η	$\varepsilon_{id}^{\mu^-}$	$\varepsilon_{id}^{\mu^+}$	ε_{id}^{μ}
3.0−3.5	(98.5 ± 0.7)%	$1_{-0.004}^{+0}$	(99.3 ± 0.4)%
3.5−4.0	(97.6 ± 1.2)%	(97.9 ± 1.2)%	(97.8 ± 0.8)%
4.0−4.5	(98.8 ± 1.2)%	(95.8 ± 2.4)%	(97.4 ± 1.3)%

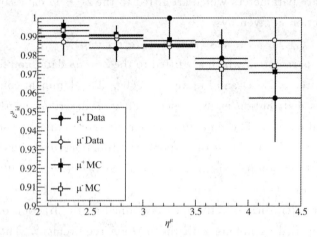

Figure 5.28 Muon identification efficiencies calculated from data and simulation

5.7.2 Muon track reconstruction efficiency

The muon track reconstruction efficiency, $\varepsilon_{track}^{\mu}$, is measured with the tag-and-probe method[89]. Tag tracks are defined as long tracks which are reconstructed from the W2Mu stripping line and are identified as muons with IsMuon=1. In order to make sure that the tag track is not a ghost track, it is required that the tag track χ^2 probability should be greater than 0.01 and its relative momentum resolution σ_p/P should be smaller than 0.1. It is also required that tag tracks should fire the trigger lines and their transverse momenta should be greater than 20 GeV/c. Probe tracks are reconstructed with hits in the muon system and TT detector. Probe tracks have opposite charges to tag tracks.

Figure. 5.29 is a schematic plot of the tag-and-probe method for the muon track reconstruction efficiency determination. As described in section 4.1.2, long tracks are reconstructed with hits in the VELO and T stations only. As TT hits are now added, the long track fit is biased. But the pattern recognition of long tracks is not affected, as TT hits are not utilized in this recognition. As a result, the bias is minimal. As these long tracks are minimally biased, we can associate muon-TT tracks to these long tracks in order to work out the tracking efficiency. The requirement applied on the muon-TT probe track is that its transverse momentum is greater than 20GeV/c and its pseudo-rapidity is in the range of $2.0 < \eta < 4.5$. The requirement on the tag and probe track combination is that its invariant mass is in the range $61 \text{ GeV}/c^2 < M_{\text{tag-probe}} < 121 \text{ GeV}/c^2$ and its dimuon acoplanarity[106] is smaller than 2.5 radians.

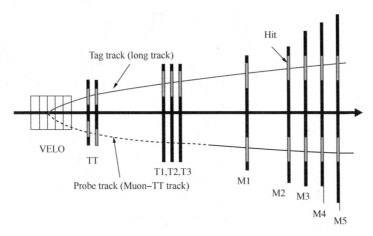

Figure 5.29　A schematic plot of the tag-and-probe method for the muon track reconstruction efficiency determination. The solid (dash) curve is a tag (probe) track. The hits are shown are grey areas

Figure. 5.30 shows the invariant mass distributions of the tag and probe tracks combinations. The left (right) plot is for the case with the tag on μ^+ (μ^-) track and the probe on the μ^- (μ^+) track. The mass distribution is fitted with the crystal ball function (see Eq. (5.12)) plus the exponential function (see Eq. (5.13)). The data sample is shown as points with error bars. The solid (dashed) line is the crystal ball (exponential) function. The momentum resolution of the muon-TT probe track is relatively poor. This is reflected by the large width of the Z peak ($\sigma_{\text{tag }\mu^+,\text{probe }\mu^-} = 10.8 \text{ GeV}/c^2$, $\sigma_{\text{tag }\mu^-,\text{probe }\mu^+} = 9.8 \text{ GeV}/c^2$).

The muon track reconstruction efficiency is defined as the fraction of probe tracks which have an associated long track in the event❶. If there are TT hits in a long track, then a muon-TT probe track is associated to that long track if at least 40% of muon chamber hits and 60% of TT hits of that probe muon-TT track are in common with that long track. If there is no TT hit in a long track, then a mu-

---

❶　It is required that the associated long track's χ^2 probability is greater than 1% and it relative momentum resolution σ_P/P is smaller than 0.1.

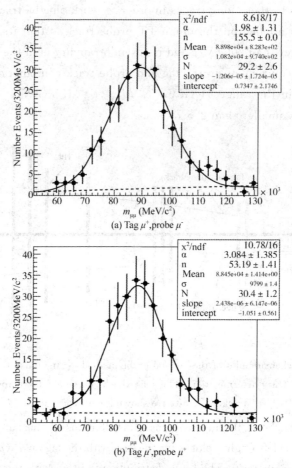

Figure 5.30 Invariant mass distributions of the tag and probe track combinations

on-TT track is associated to that long track only if at least 40% of muon chamber hits of that probe muon-TT track are in common with that long track. The track reconstruction efficiencies in five η bins for positive muons, $\varepsilon_{\text{track}}^{\mu^+}$, and negative muons, $\varepsilon_{\text{track}}^{\mu^-}$, are tabulated in Table 5.7 and they are shown as circle dots with error bars in Figure. 5.31. No charge bias is found between these two efficiencies. Thus these two efficiencies are combined together as $\varepsilon_{\text{track}}^{\mu}$ (see the fourth column of Table 5.7). Only $\varepsilon_{\text{track}}^{\mu}$ is utilized in the W cross-section calculation. In order to make a comparison with simulation, another set of tracking efficiencies are evaluated with a $Z \rightarrow \mu\mu$ simulation sample at truth level. In this sample, one muon is reconstructed in the event, the other muon is checked whether it is reconstructed to be a long track with a χ^2 probability greater than 1‰ and a relative momentum resolution σ_p/P smaller than 0.1. As the relative amount of badly reconstructed tracks in simulation is less than the relative amount of badly reconstructed tracks in data (see Figure 5.7(b)), the tracking efficiencies in simulation are higher than the efficiencies in data.

Table 5.7 Track reconstruction efficiencies in five η bins.
The uncertainties quoted are binomial errors

η	$\varepsilon_{track}^{\mu^-}$	$\varepsilon_{track}^{\mu^+}$	$\varepsilon_{track}^{\mu}$
2.0—2.5	$(79.2 \pm 3.3)\%$	$(79.9 \pm 3.0)\%$	$(79.6 \pm 2.2)\%$
2.5—3.0	$(85.5 \pm 2.7)\%$	$(82.5 \pm 2.8)\%$	$(83.9 \pm 2.0)\%$
3.0—3.5	$(91.5 \pm 2.3)\%$	$(94.0 \pm 1.8)\%$	$(92.8 \pm 1.5)\%$
3.5—4.0	$(92.0 \pm 3.1)\%$	$(91.4 \pm 3.5)\%$	$(92.0 \pm 2.3)\%$
4.0—4.5	$(92.9 \pm 4.9)\%$	$(95.7 \pm 4.3)\%$	$(94.1 \pm 3.3)\%$

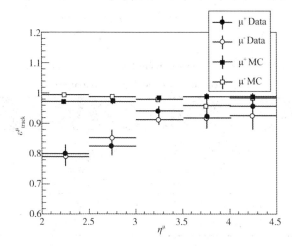

Figure 5.31 Muon track reconstruction efficiencies calculated from data and simulation

5.7.3 Muon track trigger efficiency

The muon track trigger efficiency for the $W \to \mu \nu_\mu$ process, ε_{trig}^μ, is determined as a product of two factors: one factor is the efficiency due to firing trigger lines, $\varepsilon_{trig\ lines}^\mu$; the other factor is the efficiency due to global event cuts, ε_{GEC}^μ.

A $Z \to \mu\mu$ data sample is utilized to calculate $\varepsilon_{trig\ lines}^\mu$. This sample is produced from the Z02MuMu stripping line. In the $Z \to \mu\mu$ data sample, it is required that each muon in the dimuon pair from the Z boson is greater than 20 GeV/c. It is also required that the invariant mass of the dimuon pair is in the range of 81 GeV/$c^2 < m_{\mu\mu} < 101$ GeV/c^2. Additionally the track pre-selection requirements have been applied to muon candidate tracks in the $Z \to \mu\mu$ data sample.

$\varepsilon^{\mu}_{\text{trig lines}}$ is measured with the tag-and-probe method[89]. The tag track is defined as the muon track which fires the trigger lines. The probe track is another muon track from the dimuon pair of the Z boson. $\varepsilon^{\mu}_{\text{trig lines}}$ then is calculated as the fraction of probe tracks which fire the trigger lines. The efficiencies of the positive (negative) probe muon firing the trigger lines in five η bins, $\varepsilon^{\mu^+}_{\text{trig lines}}$ ($\varepsilon^{\mu^-}_{\text{trig lines}}$), are tabulated in Table 5.8 and they are shown as circle dots with error bars in Figure. 5.32. $\varepsilon^{\mu^+}_{\text{trig lines}}$ and $\varepsilon^{\mu^-}_{\text{trig lines}}$ are consistent within uncertainties. Thus we combine them together and get a charge unbiased efficiency, $\varepsilon^{\mu}_{\text{trig lines}}$. Only $\varepsilon^{\mu}_{\text{trig lines}}$ is utilized in the W cross-section calculation. In order to compare data with simulation, another set of efficiencies for muons to fire trigger lines (square dots with error bars) are calculated with a $Z \to \mu\mu$ simulation sample. The behaviour in the data sample is not so compatible with the simulation sample in the first, second, fourth and fifth η bins. It is assumed that this discrepancy between the trigger efficiencies calculated with data and simulation is due to the imperfect trigger emulation in the simulation.

Table 5.8 Efficiencies of the probe muon track firing trigger lines in five η bins. The uncertainties quoted are binomial errors

η	$\varepsilon^{\mu}_{\text{trig lines}}$	$\varepsilon^{\mu^+}_{\text{trig lines}}$	$\varepsilon^{\mu^-}_{\text{trig lines}}$
2.0−2.5	(83.6 ± 2.2)%	(91.6 ± 2.2)%	(82.5 ± 1.5)%
2.5−3.0	(86.2 ± 2.1)%	(83.2 ± 2.1)%	(84.6 ± 1.5)%
3.0−3.5	(78.3 ± 2.5)%	(79.1 ± 2.5)%	(78.7 ± 1.8)%
3.5−4.0	(72.7 ± 3.5)%	(72.7 ± 3.8)%	(72.7 ± 2.6)%
4.0−4.5	(63.9 ± 5.3)%	(61.8 ± 5.9)%	(62.9 ± 3.9)%

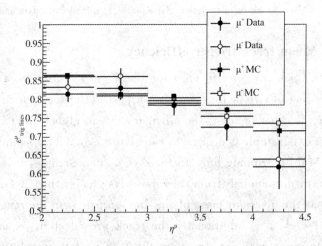

Figure 5.32 Efficiencies for muons to fire trigger lines in data and simulation

In order to calculate the global event cut efficiency for the $W \to \mu \nu_\mu$ process, the Pseudo-W data sample described in section 5.3 has been used. Since the global event cuts (GECs) are applied on the Pseudo-W data sample event by event, the GECs efficiency can not be determined by the tag-and-probe method[89].

Figure. 5.33 shows the distributions of the number of VELO clusters, the number of clusters in the IT station, the number of clusters in the OT station and the SPD multiplicity for the Pseudo-W data sample. The vertical line in each subplot shows the cuts applied in the GEC. Judging from these subplots only the GEC with the number of VELO clusters on the events in the Pseudo-W data sample is inefficient.

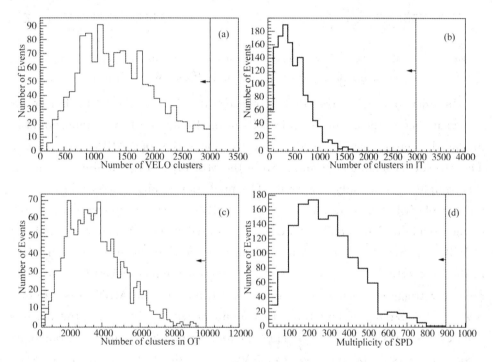

Figure 5.33 (a) Distributions of the number of VELO clusters, (b) the number of clusters in the IT station, (c) the number of clusters in the OT station, and (d) the multiplicity of SPD for the Pseudo-W data sample

Figure. 5.34(a) shows the number of VELO clusters (NVC) distribution for events with only one primary vertex in the Pseudo-W data sample. The NVC is below 3000 for every event. It can be assumed that the GEC efficiency on events with one primary vertex is about 100%. Figure. 5.34(b) shows that NVC distribution for events with three primary vertices in the Pseudo-W data sample.

It is clear that there are some events with NVC above 3000 before the application of GECs. The conclusion is that the GEC efficiency is not 100% for events with multiple PVs in the Pseudo-W data sample.

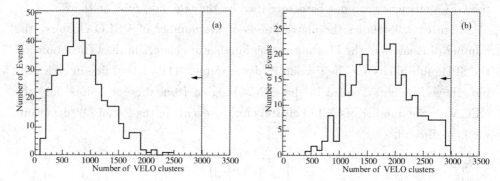

Figure 5.34 The number of VELO clusters distribution for events with (a) only one primary vertex and (b) 3 primary vertices in the Pseudo-W data sample

In order to calculate the GECs efficiency, the shape of the NVC distribution for events before the application of GECs in the Pseudo-W data sample should be predicted. This prediction is done with the application of a MBNoBias sample. The MBNoBias sample is produced from the MBNoBias stripping line. Events in this sample are randomly triggered. The prediction is obtained as follows. The number of VELO clusters for events with N PVs before the application of GECs in the Pseudo-W data sample, $\mathrm{NVC}_{N\mathrm{PVs}}^{\mathrm{BeforeGECs}}$, is predicted as the sum of the number of VELO clusters for events with 1 primary vertex after the application of GECs in the Pseudo-W data sample, $\mathrm{NVC}_{1\mathrm{PVs}}^{\mathrm{AfterGECs}}$, and the number of VELO clusters for randomly triggered events containing ($N-1$) PVs in the MBNoBias sample, $\mathrm{NVC}_{(N-1)\mathrm{PVs}}^{\mathrm{MBNoBias}}$. Simply adding these two numbers will double count the clusters from events containing 0 vertex. Thus the number of VELO clusters in randomly triggered events containing 0 vertex, $\mathrm{NVC}_{0\mathrm{PV}}^{\mathrm{MBNoBias}}$, should be subtracted in order to get the correct number of clusters in events containing N PVs. The method to do the prediction can be shown in a more clear way as follows:

$$\mathrm{NVC}_{N\mathrm{PVs}}^{\mathrm{BeforeGECs}} = \mathrm{NVC}_{1\mathrm{PVs}}^{\mathrm{AfterGECs}} + \mathrm{NVC}_{(N-1)\mathrm{PVs}}^{\mathrm{MBNoBias}} - \mathrm{NVC}_{0\mathrm{PV}}^{\mathrm{MBNoBias}} \qquad (5.14)$$

Figure. 5.35 shows the predicted NVC distributions for events containing up to 7 PVs before the application of GECs and the NVC distributions for events with the same number of PVs after the application of GECs in the Pseudo-W data sample. The agreement between the predicted NVC distribution before the application of GECs and the NVC distribution after the application of GECs for the

Pseudo-W data sample in the region with NVC$<$ 3000 confirms the hypothesis described in Eq. (5.14).

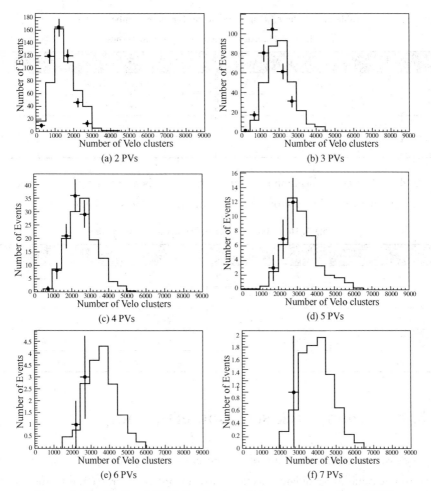

Figure 5.35 Predicted NVC distributions before the application of GECs (the histogram with the solid line) and NVC distributions after the application of GECs (points with error bars) for events containing 2 to 7 PVs in the Pseudo-W data sample

The efficiency due to GECs for NPVs ($\varepsilon_{\text{GEC}}^{\text{NPVs}}$) is the integral of the predicted NVC distribution with the NVC smaller than 3000 divided by the integral of the total predicted NVC distribution for the Pseudo-W data sample. The overall efficiency due to GECs is determined as a weighted average of the efficiencies with different PVs. The weighting for each $\varepsilon_{\text{GEC}}^{\text{NPVs}}$ is the number of events containing N PVs divided by the number of events containing all PVs in the Pseudo-W data sample before the application of GECs. The $\varepsilon_{\text{GEC}}^{\text{NPVs}}$ and weighting for each number of PVs are tabulated in Table 5.9. The total GECs efficiency for the $W \to \mu \nu_\mu$ process is $\varepsilon_{\text{GECs}} = (91.1 \pm 0.6)\%$.

Table 5.9 The $\varepsilon_{\text{GEC}}^{\text{NPVs}}$ and weighting for each number of PVs. The uncertainties quoted are binomial errors

PVs	$\varepsilon_{\text{GEC}}^{\text{NPVs}}$	Weighting	PVs	$\varepsilon_{\text{GEC}}^{\text{NPVs}}$	Weighting
1	1 ± 0	30.1%	5	$(45.8 \pm 7.2)\%$	3.3%
2	$(98.4 \pm 0.6)\%$	32.9%	6	$(23.5 \pm 10.3)\%$	1.2%
3	$(89.4 \pm 1.7)\%$	22.7%	7	$(11.3 \pm 10.6)\%$	0.1%
4	$(71.3 \pm 3.9)\%$	9.2%			

As described at the beginning of this section, $\varepsilon_{\text{trig}}^\mu$ is a product of $\varepsilon_{\text{trig lines}}^\mu$ and $\varepsilon_{\text{GECs}}$, thus we combine these two factors and tabulate the products in Table 5.10.

Table 5.10 Muon trigger efficiencies in five η bins

η	$\varepsilon_{\text{trig}}^{\mu^-}$	$\varepsilon_{\text{trig}}^{\mu^+}$	$\varepsilon_{\text{trig}}^{\mu^\pm}$
2.0 − 2.5	$(76.2 \pm 2.1)\%$	$(74.3 \pm 2.1)\%$	$(75.2 \pm 1.5)\%$
2.5 − 3.0	$(78.5 \pm 2.0)\%$	$(75.8 \pm 2.0)\%$	$(77.1 \pm 1.5)\%$
3.0 − 3.5	$(71.3 \pm 2.3)\%$	$(72.1 \pm 2.3)\%$	$(71.7 \pm 1.7)\%$
3.5 − 4.0	$(66.2 \pm 3.2)\%$	$(66.2 \pm 3.5)\%$	$(66.2 \pm 2.4)\%$
4.0 − 4.5	$(58.2 \pm 4.8)\%$	$(56.3 \pm 5.4)\%$	$(57.3 \pm 3.6)\%$

5.8 Muon track selection efficiency

As described at the beginning of section 5.5, the real number of signal events is underestimated due to candidate selection cuts. In order to correct the inefficiency, the selection efficiency is applied. The candidate selection cuts applied on the muon candidates with 20 GeV/c $< p_T^\mu <$ 70 GeV/c are listed in Table 5.11.

Table 5.11 Candidate selection cuts applied on the muon candidates with 20 GeV/c $< p_T^\mu <$ 70 GeV/c

Variables	Requirements	Variables	Requirements
p_T^{Extra}	<2 GeV/c	"Charged" p_T^{cone}	<2 GeV/c
Unbiased IP	<0.04 mm	"Neutral" p_T^{cone}	<2 GeV/c
$E_{\text{ECAL+HCAL}}/pc$	<0.04		

Since the $N_{\text{TThits}} > 0$ requirement is not taken into consideration when the muon track reconstruction efficiency is calculated, the efficiency for the $N_{\text{TThits}} > 0$ cut is included in this selection efficiency. As described in section 5.4, the Pseudo-W data sample is in general agreement with the $W \to \mu \nu_\mu$ simulation sample. Thus this data sample is utilized to determine the selection efficiency, $\varepsilon_{\text{selection}}^{\text{Pseudo-W Data}}$ (see the second column of Table 5.12). No charge bias is observed between selection efficiencies for positive and negative muons. The uncertainties quoted on $\varepsilon_{\text{selection}}^{\text{Pseudo-W Data}}$ are binomial errors. However, as the p_T spectrum is harder in the $Z \to \mu\mu$ decay than in the $W \to \mu \nu_\mu$ decay, a difference is observed in the relative energy deposition distributions of the Pseudo-W data sample and $W \to \mu \nu_\mu$ simulation sample. As a result, the selection efficiency calculated with the Pseudo-W data sample is different to the efficiency calculated with the $W \to \mu \nu_\mu$ simulation sample. This selection efficiency difference, $\Delta\varepsilon_{\text{selection}}^{\text{MC}}$ (see third column of Table 5.12), is estimated with $W \to \mu \nu_\mu$ and $Z \to \mu\mu$ simulation samples and it is utilized to correct the selection efficiency calculated from the Pseudo-W data sample. The uncertainties quoted on $\Delta\varepsilon_{\text{selection}}^{\text{MC}}$ is evaluated by adding binomial uncertainties of efficiencies calculated from these two simulation samples in quadrature. The fourth column of Table 5.12 shows the corrected selection efficiency for the Pseudo-W data sample in each η bin, $\varepsilon_{\text{selection}}^{\mu}$. The uncertainty quoted on $\varepsilon_{\text{selection}}^{\mu}$ is determined by adding uncertainties on $\Delta\varepsilon_{\text{selection}}^{\text{Pseudo-W Data}}$ and $\Delta\varepsilon_{\text{selection}}^{\text{MC}}$ in quadrature.

Table 5.12 Selection efficiencies calculated from the Pseudo-W data sample (the second column), differences between selection efficiencies calculated from the $W \to \mu \nu_\mu$ and $Z \to \mu\mu$ simulation samples (the third column) and corrected selection efficiencies for the $W \to \mu \nu_\mu$ process (the fourth column) in five η bins

η	$\varepsilon_{\text{selection}}^{\text{Pseudo-W Data}}$	$\Delta\varepsilon_{\text{selection}}^{\text{MC}}$	$\varepsilon_{\text{selection}}^{\mu}$
2.0 − 2.5	$(52.9 \pm 1.6)\%$	$(-1.2 \pm 0.6)\%$	$(51.7 \pm 1.7)\%$
2.5 − 3.0	$(65.6 \pm 1.7)\%$	$(-0.9 \pm 0.6)\%$	$(64.7 \pm 1.8)\%$
3.0 − 3.5	$(68.9 \pm 1.8)\%$	$(-0.6 \pm 0.8)\%$	$(68.3 \pm 1.9)\%$
3.5 − 4.0	$(71.4 \pm 2.3)\%$	$(-2.6 \pm 1.0)\%$	$(68.8 \pm 2.5)\%$
4.0 − 4.5	$(33.2 \pm 3.5)\%$	$(1.9 \pm 1.7)\%$	$(35.1 \pm 3.9)\%$

5.9 Muon track acceptance efficiency

As described in the beginning of section 5.5, in order to calculate the $W \to \mu$

ν_μ cross-section in the fiducial phase-space with the muon p_T greater than 20 GeV/c, the number of signal events in the muon p_T range between 20GeV/c and 70 GeV/c has to be corrected with the acceptance efficiency. This efficiency is defined as the number of signal events with 20 GeV/c$<p_T<$70 GeV/c divided by the number of signal events with $p_T>$20 GeV/c. Since there are not enough statistics with $p_T>$70 GeV/c and events only contain LO information in the $W \to \mu \nu_\mu$ simulation sample, the POWHEG simulation sample is utilized instead to calculate the acceptance efficiency. Table 5.13 shows the acceptance efficiency in five η bins for the positive muon, $\varepsilon_{acceptance}^{\mu^+}$, and for the negative muon, $\varepsilon_{acceptance}^{\mu^-}$. The charge unbiased acceptance efficiencies, $\varepsilon_{acceptance}^{\mu}$, in five eta bins are shown in the fourth column of Table 5.13. Only $\varepsilon_{acceptance}^{\mu}$ is utilized in the W cross-section calculation. The application of the fiducial phase-space cut is considered as a binomial process, thus the uncertainties quoted in Table 5.13 are binomial errors. In the last η bin, the acceptance efficiencies are determined by inverting the probability in the binomial process with Bayes' Theorem.

Table 5.13 Acceptance efficiencies for the $W \to \mu \nu_\mu$ process in five η bins

η	$\varepsilon_{acceptance}^{\mu^+}$	$\varepsilon_{acceptance}^{\mu^-}$	$\varepsilon_{acceptance}^{\mu}$
2.0 − 2.5	(99.19 ± 0.04)%	(98.50 ± 0.05)%	(98.88 ± 0.03)%
2.5 − 3.0	(99.52 ± 0.03)%	(99.08 ± 0.05)%	(99.30 ± 0.03)%
3.0 − 3.5	(99.78 ± 0.03)%	(99.58 ± 0.03)%	(99.66 ± 0.02)%
3.5 − 4.0	(99.94 ± 0.03)%	(99.90 ± 0.02)%	(99.91 ± 0.02)%
4.0 − 4.5	$1_{-0.0005}^{+0}$	$1_{-0.0001}^{+0}$	$1_{-0.0001}^{+0}$

5.10 Final state radiation

As described at the beginning of section 5.5, in order to make a consistent comparison with theoretical predictions, where the QED FSR is not present, the W cross-section measurements in data are corrected to Born level. In order to estimate the FSR corrections, the PHOTOS generator[107], interfaced to the PYTHIA generator, is utilized. The FSR correction factor is given by the difference of cross-sections estimated by PYTHIA (pre-FSR) and PHOTOS (post-FSR). The p_T spectrum of W bosons in PHOTOS has been reweighted to the NNLO W boson p_T distribution generated by DYNNLO[108,109]. Table 5.14 shows the FSR correction factors, σ_{FSR}^W, in each η bin as well as in the whole η range for the W^+ and W^-

cross-sections. The uncertainty quoted in the table is the maximum of statistical uncertainties on the corrections with re-weighting and without re-weighting the W boson p_T to NNLO.

Table 5.14 FSR correction factors in each η bin as well as in the whole η range for the W^+ and W^- cross-sections

η	$\sigma_{FSR}^{W^+}$ (pb)	$\sigma_{FSR}^{W^-}$ (pb)
2.0 −2.5	5.05± 0.15	2.89± 0.16
2.5 −3.0	2.01± 0.05	3.01± 0.07
3.0 −3.5	1.59± 0.09	2.07± 0.05
3.5 −4.0	0.58± 0.03	1.73± 0.08
4.0 −4.5	0.01± 0.01	0.66± 0.03
2.0 −4.5	9.24± 0.14	10.36± 0.13

5.11 $\sigma_{W \to \mu \nu_\mu}$ determination

5.11.1 Differential cross-section determination

The cross-section of the W boson decaying into a muon and a neutrino is measured in the fiducial phase-space with $p_T^\mu > 20$ GeV/c and $2.0 < \eta^\mu < 4.5$.

As described at the beginning of section 5.5, the cross-section in each pseudo-rapidity bin, η_i, is calculated with the following formula:

$$\sigma_{W \to \mu \nu_\mu}(\eta_i) = \frac{N_{obs}^W(\eta_i) - N_{bkg}^W(\eta_i)}{\int L \; \varepsilon_{detector}^\mu(\eta_i) \cdot \varepsilon_{selection}^\mu(\eta_i) \cdot \varepsilon_{acceptance}^\mu(\eta_i)} + \sigma_{bkg}^W(\eta_i) \quad (5.15)$$

where $\int L$ is the integrated luminosity of the data sample used in this analysis, $\int L = 37.1 \pm 1.3$ pb^{-1}. $N_{obs}^W(\eta_i)$ is the number of observed $W \to \mu \nu_\mu$ candidates in a given η bin (see Table 5.4). $N_{bkg}^W(\eta_i)$ is the estimated number of background events in a given η bin (see Table 5.5). $\varepsilon_{detector}^\mu(\eta_i)$ is the overall detector efficiency which is a product of the muon charge unbiased track identification efficiency, ε_{id}^μ (see Table 5.6), the muon charge unbiased track reconstruction efficiency, ε_{track}^μ (see Table 5.7) and the muon charge unbiased track trigger efficiency, ε_{trig}^μ (see Table 5.10) in a given η bin. $\varepsilon_{selection}^\mu(\eta_i)$ is the muon charge unbiased track selection efficiency in a given η bin (see Table 5.12). $\varepsilon_{acceptance}^\mu(\eta_i)$ is the muon charge unbiased track acceptance efficiency in a given η bin (see Table 5.13). σ_{FSR}^W is the

QED FSR correction to the W cross-section in a given η bin (see Table 5.14). The total cross-section then is determined by summing over the cross-sections in each η bin:

$$\sigma_{W \to \mu \nu_\mu}(p_T^\mu > 20 \text{ GeV}/c, 2.0 < \eta^\mu < 4.5) = \sum_{\eta_i} \sigma_{W \to \mu \nu_\mu}(\eta_i) \quad (5.16)$$

Table 5.15 shows the differential as well as total cross-sections for the $W^+ \to \mu^+ \nu_\mu$ and $W^- \to \mu^- \bar{\nu}_\mu$ processes. In this table, the first uncertainty is the statistical uncertainty, the second is the systematic uncertainty and the third is due to the luminosity determination.

Table 5.15 The differential as well as total cross-sections for the $W^+ \to \mu^+ \nu_\mu$ and $W^- \to \mu^- \bar{\nu}_\mu$ processes

η	$\sigma_{W^+ \to \mu^+ \nu_\mu}$ (pb)	$\sigma_{W^- \to \mu^- \bar{\nu}_\mu}$ (pb)
2.0 — 2.5	390.8 ± 6.6 ± 19.3 ± 13.5	217.3 ± 5.2 ± 11.6 ± 7.5
2.5 — 3.0	280.4 ± 4.7 ± 11.9 ± 9.8	195.0 ± 4.0 ± 8.5 ± 6.7
3.0 — 3.5	149.3 ± 3.4 ± 6.4 ± 5.2	143.8 ± 3.2 ± 6.0 ± 5.0
3.5 — 4.0	58.5 ± 2.2 ± 3.7 ± 2.0	98.6 ± 2.7 ± 5.6 ± 3.4
4.0 — 4.5	11.1 ± 1.6 ± 2.2 ± 0.4	32.3 ± 2.3 ± 4.3 ± 1.1
2.0 — 4.5	890.1 ± 9.2 ± 26.4 ± 30.9	687.0 ± 8.1 ± 19.6 ± 23.7

5.11.2 Statistical uncertainty on the cross-section

The statistical uncertainty on the cross-section in each η bin, $\Delta \sigma_{\text{stat.}}(\eta_i)$, is calculated as follows:

$$\Delta \sigma_{\text{stat.}}(\eta_i) = \frac{\Delta N_{\text{obs}}^W(\eta_i)}{\int L \, \varepsilon_{\text{detector}}^\mu(\eta_i) \cdot \varepsilon_{\text{selection}}^\mu(\eta_i) \cdot \varepsilon_{\text{acceptance}}^\mu(\eta_i)} \quad (5.17)$$

where $\Delta N_{\text{obs}}^W(\eta_i)$ is the statistical uncertainty on the number of observed events, and is determined as the square root of the number of observed events in the data sample (see Table 5.4). The statistical uncertainties on the differential cross-sections are shown in Table 5.15. The statistical uncertainties on the total cross-sections are determined by summing over the statistical uncertainties in each η bin in quadrature.

5.11.3 Systematic uncertainty on the cross-section

There are several sources for the systematic uncertainties: the background estimation; the background template normalization; the detector efficiency determination; the selection efficiency determination; the acceptance efficiency determination; the luminosity determination and the FSR correction. Each of these sources now is described in detail.

Systematic uncertainty from the background estimation

The systematic uncertainty on the cross-section due to the background estimation in a given η bin, $\Delta\sigma_{bkg}(\eta_i)$, is calculated as follows:

$$\Delta\sigma_{bkg}(\eta_i) = \frac{\Delta N^W_{bkg}(\eta_i)}{\int L\, \varepsilon^\mu_{detector}(\eta_i) \cdot \varepsilon^\mu_{selection}(\eta_i) \cdot \varepsilon^\mu_{acceptance}(\eta_i)} \quad (5.18)$$

where $\Delta N^W_{bkg}(\eta_i)$ is the statistical uncertainty on the number of estimated background events in a given η bin, as returned by the p_T spectrum fit (see Table 5.5). The systematic uncertainties from the background estimation on the cross-sections for the $W^+ \to \mu^+ \nu_\mu$ and $W^- \to \mu^- \bar{\nu}_\mu$ processes in each η bin as well as in the whole η range are shown in Table 5.16. As described in section 5.6, the statistical uncertainties on the number of estimated background events are correlated between η bins, thus the systematic uncertainties on the total cross-sections due to the background estimation are determined by summing over the uncertainties in each η bin linearly.

Table 5.16 Systematic uncertainties from the background estimation on the differential and total cross-section for the $W^+ \to \mu^+ \nu_\mu$ and $W^- \to \mu^- \bar{\nu}_\mu$ processes

η	$\Delta\sigma_{W^+ \to \mu^+ \nu_\mu}$ (pb)	$\Delta\sigma_{W^- \to \mu^- \bar{\nu}_\mu}$ (pb)
2.0 −2.5	±3.2	±2.6
2.5 −3.0	±2.2	±1.4
3.0 −3.5	±2.1	±1.2
3.5 −4.0	±1.5	±0.8
4.0 −4.5	±1.6	±0.7
2.0 −4.5	±10.5	±6.7

Systematic uncertainty from the background template normalization

The systematic uncertainty on the cross-section due to the background template normalization is determined by varying the fractions of two main background templates ($Z \to \mu\mu$ and $W \to \tau\nu$) by $\pm 1\sigma$. After each background template fraction variation, the p_T spectrum fit is performed again to determine the new fractions of signal and background templates. Then new differential and total

W cross-sections are determined. Finally, the systematic uncertainty due to this background template fraction variation (see Table 5.15) is determined as the largest difference between the new and original cross-sections.

The fraction of the $Z \to \mu\mu$ template is varied by $\pm 0.5\%$ ❶ and $\pm 0.9\%$ ❷ separately (see section 5.5.5). Table 5.17 shows differences between the new fractions after these two template fraction variations and original fractions in Table 5.3. The observed differences on the differential and total W cross-sections are determined by adding the differences due to these two template fraction variations in quadrature (see Table 5.18). It is assumed that the differences on the cross-sections in each η bin are correlated, thus the differences on the total cross-sections are determined by adding the differences in each η bin linearly.

Table 5.17 Differences between the new fractions after the $Z \to \mu\mu$ template fraction variations and original fractions in Table 5.3. The second and third columns show the differences due to the $\pm 0.9\%$ variation. The fourth and fifth columns show the differences due to the $\pm 0.5\%$ variation

Sample	$\Delta\%(+1\text{syst.}\sigma)$	$\Delta\%(-1\text{syst.}\sigma)$	$\Delta\%(+1\text{stat.}\sigma)$	$\Delta\%(-1\text{stat.}\sigma)$
$W^+ \to \mu^+ \nu_\mu$	-0.32%	+0.32%	-0.18%	+0.17%
$W^- \to \mu^- \bar{\nu}_\mu$	-0.40%	+0.41%	-0.22%	+0.22%
Decay in flight	-0.15%	+0.13%	-0.07%	+0.09%
$Z \to \mu\mu$	-0.90%	+0.90%	-0.50%	+0.50%

Table 5.18 Observed differences on the differential and total W cross-sections due to the $Z \to \mu\mu$ template fraction variations. The differences in brackets are due to the -1σ variation

η	$\Delta\sigma_{W^+ \to \mu^+ \nu_\mu}$ (pb)	$\Delta\sigma_{W^- \to \mu^- \bar{\nu}_\mu}$ (pb)
2.0 — 2.5	-4.03(+3.96)	-4.38(+4.33)
2.5 — 3.0	-2.19(+2.16)	-2.66(+2.64)
3.0 — 3.5	-0.98(+0.95)	-1.43(+1.40)
3.5 — 4.0	-0.36(+0.34)	-0.68(+0.68)
4.0 — 4.5	-0.06(+0)	-0.24(+0.21)
2.0 — 4.5	-7.61(+7.40)	-9.39(+9.26)

The fraction of the $W \to \tau\nu$ template is varied by $\pm 0.2\%$ (see section 5.5.6). Table 5.19 shows differences between the new fractions after this template fraction variation and original fractions. The observed differences on the differential and total W

---

❶ Statistical uncertainty on the $Z \to \mu\mu$ fraction.
❷ Systematic uncertainty on the $Z \to \mu\mu$ fraction.

cross-sections due to this template fraction variation are tabulated in Table 5.20.

Table 5.19 Differences between the new fractions after the $W \to \tau \nu$ template fraction variations and original fractions

Sample	$\Delta\%(+1\sigma)$	$\Delta\%(-1\sigma)$
$W^+ \to \mu^+ \nu_\mu$	-0.06%	+0.07%
$W^- \to \mu^- \bar{\nu}_\mu$	-0.04%	+0.13%
Decay in flight	-0.12%	+0.11%
$W \to \tau \nu$	+0.20%	-0.20%

Table 5.20 Observed differences on the differential and total W cross-sections due to the $W \to \tau \nu$ template fraction variation. The differences in brackets are due to the -1σ variation

η	$\Delta\sigma_{W^+ \to \mu^+ \nu_\mu}$ (pb)	$\Delta\sigma_{W^- \to \mu^- \bar{\nu}_\mu}$ (pb)
2.0 — 2.5	-0.73 (+0.71)	-0.21 (+0.20)
2.5 — 3.0	-0.47 (+0.47)	-0.15 (+0.15)
3.0 — 3.5	-0.26 (+0.43)	-0.20 (+0.29)
3.5 — 4.0	-0.07 (+0.05)	-0.11 (+0.10)
4.0 — 4.5	-0.06 (+0.02)	-0.09 (+0.08)
2.0 — 4.5	-1.59 (+1.68)	-0.76 (+0.82)

It is assumed that systematic uncertainties due to each background template fraction variation are uncorrelated, thus the systematic uncertainty in each η bin due to these two template fraction variations is determined by adding the largest systematic uncertainty from each variation in quadrature. The systematic uncertainty due to these two template fraction variations in the whole η range then is determined by adding the systematic uncertainty in each η bin linearly (see Table 5.21).

Table 5.21 Total systematic uncertainties in each η bin as well as the whole η range due to background template fraction variations

η	$\Delta\sigma_{W^+ \to \mu^+ \nu_\mu}$ (pb)	$\Delta\sigma_{W^- \to \mu^- \bar{\nu}_\mu}$ (pb)
2.0 — 2.5	±4.10	±4.39
2.5 — 3.0	±2.24	±2.66
3.0 — 3.5	±1.07	±1.46
3.5 — 4.0	±0.37	±0.69
4.0 — 4.5	±0.08	±0.26
2.0 — 4.5	±7.79	±9.43

Systematic uncertainty from the detector efficiency determination

The systematic uncertainty on the cross-section due to the muon track identification (reconstruction, trigger) efficiency determination in a given η bin $\Delta\sigma_{\varepsilon^\mu_{id}}(\eta_i)(\Delta\sigma_{\varepsilon^\mu_{track}}(\eta_i),\Delta\sigma_{\varepsilon^\mu_{trig}}(\eta_i))$, is calculated as follows[❶]:

$$\begin{cases} \Delta\sigma_{\varepsilon^\mu_{id}}(\eta_i) = \dfrac{(N^W_{obs}(\eta_i)-N^W_{bkg}(\eta_i))\cdot\Delta\varepsilon^\mu_{id}(\eta_i)}{\int L \cdot \varepsilon^\mu_{id}(\eta_i)\cdot\varepsilon^\mu_{detector}(\eta_i)\cdot\varepsilon^\mu_{selection}(\eta_i)\cdot\varepsilon^\mu_{acceptance}(\eta_i)} \\[2mm] \Delta\sigma_{\varepsilon^\mu_{track}}(\eta_i) = \dfrac{(N^W_{obs}(\eta_i)-N^W_{bkg}(\eta_i))\cdot\Delta\varepsilon^\mu_{track}(\eta_i)}{\int L \cdot \varepsilon^\mu_{track}(\eta_i)\cdot\varepsilon^\mu_{detector}(\eta_i)\cdot\varepsilon^\mu_{selection}(\eta_i)\cdot\varepsilon^\mu_{acceptance}(\eta_i)} \\[2mm] \Delta\sigma_{\varepsilon^\mu_{trig}}(\eta_i) = \dfrac{(N^W_{obs}(\eta_i)-N^W_{bkg}(\eta_i))\cdot\Delta\varepsilon^\mu_{trig}(\eta_i)}{\int L \cdot \varepsilon^\mu_{trig}(\eta_i)\cdot\varepsilon^\mu_{detector}(\eta_i)\cdot\varepsilon^\mu_{selection}(\eta_i)\cdot\varepsilon^\mu_{acceptance}(\eta_i)} \end{cases}$$

(5.19)

where $\Delta\varepsilon^\mu_{id}(\eta_i)(\Delta\varepsilon^\mu_{track}(\eta_i),\Delta\varepsilon^\mu_{trig}(\eta_i))$ is the uncertainty of $\varepsilon^\mu_{id}(\eta_i)$ ($\varepsilon^\mu_{track}(\eta_i)$, $\varepsilon^\mu_{trig}(\eta_i)$) in a given η bin (see Tables 5.6, 5.7 and 5.10).

Since $\Delta\varepsilon^\mu_{id}(\eta_i)$, $\Delta\varepsilon^\mu_{track}(\eta_i)$ and $\Delta\varepsilon^\mu_{trig}(\eta_i)$ are uncorrelated with each other, and $\varepsilon^\mu_{detector}(\eta_i)$ is a product of these three efficiencies, the systematic uncertainty from the $\varepsilon^\mu_{detector}(\eta_i)$ determination, $\Delta\sigma_{\varepsilon^\mu_{detector}}(\eta_i)$, then is determined as follows:

$$\Delta\sigma_{\varepsilon^\mu_{detector}}(\eta_i) = \sqrt{(\Delta\sigma_{\varepsilon^\mu_{id}}(\eta_i))^2+(\Delta\sigma_{\varepsilon^\mu_{track}}(\eta_i))^2+(\Delta\sigma_{\varepsilon^\mu_{trig}}(\eta_i))^2} \quad (5.20)$$

$\Delta\sigma_{\varepsilon^\mu_{detector}}$ in each η bin as well as in the whole η range are tabulated in Table 5.22. It is assumed that uncertainties on the detector efficiencies are uncorrelated between η bins, thus the systematic uncertainties on the total cross-sections due to the detector efficiency determinations are calculated by adding the uncertainties in each η bin in quadrature.

Table 5.22 Systematic uncertainties from the detector efficiency determination on the differential and total cross-sections for the $W^+ \to \mu^+\nu_\mu$ and $W^- \to \mu^-\bar{\nu}_\mu$ processes

η	$\Delta\sigma^{W^+\to\mu^+\nu_\mu}_{\varepsilon^\mu_{detector}}$ (pb)	$\Delta\sigma^{W^-\to\mu^-\bar{\nu}_\mu}_{\varepsilon^\mu_{detector}}$ (pb)
2.0 — 2.5	± 13.4	± 7.4
2.5 — 3.0	± 8.5	± 5.8
3.0 — 3.5	± 4.2	± 4.0
3.5 — 4.0	± 2.6	± 4.3
4.0 — 4.5	± 0.8	± 2.3
2.0 — 4.5	± 16.6	± 11.4

---

❶ Here only the statistical uncertainty on the muon track identification, reconstruction or trigger efficiency is considered as a source of systematic uncertainty on the cross-section. The systematic uncertainty on the efficiency due to the tag-and-probe method is not considered.

Systematic uncertainty from the selection efficiency determination

The systematic uncertainty on the cross-section due to the selection efficiency determination in a given η bin, $\Delta\sigma_{\varepsilon_{\text{selection}}^{\mu}}(\eta_i)$, is calculated as follows:

$$\Delta\sigma_{\varepsilon_{\text{selection}}^{\mu}}(\eta_i) = \frac{(N_{\text{obs}}^{W}(\eta_i) - N_{\text{bkg}}^{W}(\eta_i)) \cdot \Delta\varepsilon_{\text{selection}}^{\mu}(\eta_i)}{\int L \cdot \varepsilon_{\text{detector}}^{\mu}(\eta_i) \cdot (\varepsilon_{\text{selection}}^{\mu}(\eta_i))^2 \cdot \varepsilon_{\text{acceptance}}^{\mu}(\eta_i)} \quad (5.21)$$

where $\Delta\varepsilon_{\text{selection}}^{\mu}(\eta_i)$ is the uncertainty of $\varepsilon_{\text{selection}}^{\mu}(\eta_i)$ in a given η bin (see Table 5.12). The systematic uncertainties on the differential and total cross-sections due to the selection efficiency determinations are tabulated in Table 5.23. It is assumed that uncertainties on the selection efficiencies in each η bin are uncorrelated, thus the systematic uncertainties on the total cross-sections are determined by adding the uncertainties in each η bin in quadrature.

Table 5.23　Systematic uncertainties from the selection efficiency determination on the differential and total cross-sections for the $W^+ \to \mu^+ \nu_\mu$ and $W^- \to \mu^- \bar{\nu}_\mu$ processes

η	$\Delta\sigma_{\varepsilon_{\text{selection}}^{\mu}}^{W^+ \to \mu^+ \nu_\mu}$ (pb)	$\Delta\sigma_{\varepsilon_{\text{selection}}^{\mu}}^{W^- \to \mu^- \bar{\nu}_\mu}$ (pb)
2.0 — 2.5	± 13.0	± 7.2
2.5 — 3.0	± 7.8	± 5.4
3.0 — 3.5	± 4.2	± 4.0
3.5 — 4.0	± 2.6	± 3.5
4.0 — 4.5	± 1.2	± 3.5
2.0 — 4.5	± 16.0	± 11.4

Systematic uncertainty from the acceptance efficiency determination

The systematic uncertainty on the cross-section due to the acceptance efficiency determination in a given η bin, $\Delta\sigma_{\varepsilon_{\text{acceptance}}^{\mu}}(\eta_i)$, is calculated as follows:

$$\Delta\sigma_{\varepsilon_{\text{acceptance}}^{\mu}}(\eta_i) = \frac{(N_{\text{obs}}^{W}(\eta_i) - N_{\text{bkg}}^{W}(\eta_i)) \cdot \Delta\varepsilon_{\text{acceptance}}^{\mu}(\eta_i)}{\int L \cdot \varepsilon_{\text{detector}}^{\mu}(\eta_i) \cdot \varepsilon_{\text{selection}}^{\mu}(\eta_i) \cdot (\varepsilon_{\text{acceptance}}^{\mu}(\eta_i))^2} \quad (5.22)$$

where $\Delta\varepsilon_{\text{acceptance}}^{\mu}(\eta_i)$ is the uncertainty of $\varepsilon_{\text{acceptance}}^{\mu}(\eta_i)$ in a given η bin (see Table 5.13). The systematic uncertainties on the differential and total cross-sections due to the acceptance efficiency determinations are tabulated in Table 5.24. It is assumed that uncertainties on the acceptance efficiencies in each η bin are not correlated, thus the systematic uncertainties on the total cross-sections are calculated by adding the uncertainties in each η bin in quadrature.

Table 5.24 Systematic uncertainties from the acceptance efficiency determination on the differential and total cross-sections for the $W^+ \to \mu^+ \nu_\mu$ and $W^- \to \mu^- \bar{\nu}_\mu$ processes

η	$\Delta \sigma^{W^+ \to \mu^+ \nu_\mu}_{\epsilon^\mu_{\text{acceptance}}}$ (pb)	$\Delta \sigma^{W^- \to \mu^- \bar{\nu}_\mu}_{\epsilon^\mu_{\text{acceptance}}}$ (pb)
2.0 — 2.5	±0.117	±0.065
2.5 — 3.0	±0.084	±0.058
3.0 — 3.5	±0.030	±0.028
3.5 — 4.0	±0.012	±0.019
4.0 — 4.5	±0.001	±0.003
2.0 — 4.5	±0.148	±0.094

Systematic uncertainty from the FSR correction

The systematic uncertainty on the cross-section from the FSR correction is taken as the statistical uncertainty on the correction (see Table 5.14). This uncertainty is the maximum of statistical uncertainties on the corrections with re-weighting and without re-weighting the W boson p_T to NNLO.

Total systematic uncertainties

As systematic uncertainties from each source are assumed to be uncorrelated, the total systematic uncertainty (see Table 5.15) is determined by adding the uncertainty due to the background estimation (see Table 5.16), the uncertainty due to the detector efficiency determination (see Table 5.22), the uncertainty due to the selection determination (see Table 5.23), the uncertainty due to the acceptance efficiency determination (see Table 5.24), the uncertainty due to the background template normalization (see Table 5.21), and the uncertainty due to the FSR correction (see Table 5.14) in quadrature. Table 5.25 (5.26) shows percentage systematic uncertainties on the differential and total W^+ (W^-) cross-sections. The systematic uncertainty due to the detector efficiency determination is the largest one, except in the last η bin, where the systematic uncertainty on the W^+ (W^-) cross-section due to the background estimation (selection efficiency determination) is the largest one.

Table 5.25 Percentage systematic uncertainties on the W^+ cross-sections in each η bin as well as in the whole η range

Systematic Source	2.0-2.5	2.5-3.0	3.0-3.5	3.5-4.0	4.0-4.5	2.0-4.5
Background Estimation	0.81%	0.77%	1.38%	2.61%	14.67%	1.18%
Background Normalization	1.05%	0.80%	0.72%	0.63%	0.54%	0.88%
Detector Efficiency	3.42%	3.01%	2.81%	4.42%	7.29%	1.86%
Selection Efficiency	3.32%	2.78%	2.78%	3.52%	11.07%	1.78%

continue table

Systematic Source	2.0-2.5	2.5-3.0	3.0-3.5	3.5-4.0	4.0-4.5	2.0-4.5
Acceptance Efficiency	0.03%	0.03%	0.02%	0.02%	0.01%	0.02%
FSR correction	0.04%	0.02%	0.06%	0.05%	0.09%	0.02%
Total Systematic Uncertainty	4.94%	4.25%	4.25%	6.33%	19.8%	2.97%

Table 5.26 Percentage systematic uncertainties on the W^- cross-sections in each η bin as well as in the whole η range

Systematic Source	2.0-2.5	2.5-3.0	3.0-3.5	3.5-4.0	4.0-4.5	2.0-4.5
Background Estimation	1.21%	0.71%	0.82%	0.78%	2.26%	0.97%
Background Normalization	2.02%	1.36%	1.02%	0.70%	0.81%	1.37%
Detector Efficiency	3.41%	2.99%	2.80%	4.39%	7.16%	1.66%
Selection Efficiency	3.32%	2.75%	2.78%	3.50%	10.87%	1.60%
Acceptance Efficiency	0.03%	0.03%	0.02%	0.02%	0.01%	0.01%
FSR correction	0.03%	0.04%	0.03%	0.08%	0.09%	0.02%
Total Systematic Uncertainty	5.31%	4.35%	4.15%	5.71%	13.26%	2.85%

5.11.4 Uncertainty due to the luminosity determination

The uncertainty on the cross-section due to the luminosity determination in a given η bin, $\Delta\sigma_{\text{lumi}}(\eta_i)$, is calculated as follows:

$$\Delta\sigma_{\text{lumi}}(\eta_i) = \frac{N^W_{\text{obs}}(\eta_i) - N^W_{\text{bkg}}(\eta_i)}{\left(\int L\right)^2 \cdot \varepsilon^\mu_{\text{detector}}(\eta_i) \cdot \varepsilon^\mu_{\text{selection}}(\eta_i) \cdot \varepsilon^\mu_{\text{acceptance}}(\eta_i)} \cdot \Delta\left(\int L\right) \quad (5.23)$$

where $\Delta\left(\int L\right)$ is the uncertainty on the integrated luminosity determination, $\Delta\left(\int L\right) = 1.3 \text{ pb}^{-1}$. Uncertainties on the differential and total W cross-sections due to the luminosity determinations are shown in Table 5.15. It is assumed that uncertainties on the cross-sections in each η bin due to the luminosity determinations are correlated, thus uncertainties on the total W cross-sections are determined by adding uncertainties in each η bin linearly. As $\Delta\left(\int L\right)/\int L$ is 3.5%, the percentage uncertainties on the differential and total W cross-sections due to the luminosity determination are 3.5%.

5.12 $\sigma_{W \to \mu\nu_\mu}$ ratio

The ratio between cross-sections for the $W^+ \to \mu^+ \nu_\mu$ and $W^- \to \mu^- \bar{\nu}_\mu$ processes, R_{+-}, is determined as follows:

$$R_{+-} = \frac{\sigma_{W^+ \to \mu^+ \nu_\mu}}{\sigma_{W^- \to \mu^- \bar{\nu}_\mu}} \tag{5.24}$$

It is assumed that the statistical uncertainties on the W^+ and W^- cross-sections, $\Delta\sigma^{stat.}_{W^+ \to \mu^+ \nu_\mu}$ and $\Delta\sigma^{stat.}_{W^- \to \mu^- \bar{\nu}_\mu}$, are not correlated, thus the statistical uncertainty on R_{+-}, $\Delta R^{stat.}_{+-}$, is

$$\Delta R^{stat.}_{+-} = \sqrt{\left(\frac{\Delta\sigma^{stat.}_{W^+ \to \mu^+ \nu_\mu}}{\sigma_{W^+ \to \mu^+ \nu_\mu}}\right)^2 + \left(\frac{\Delta\sigma^{stat.}_{W^- \to \mu^- \bar{\nu}_\mu}}{\sigma_{W^- \to \mu^- \bar{\nu}_\mu}}\right)^2} \cdot R_{+-} \tag{5.25}$$

Since the FSR corrections on the W^+ and W^- cross-sections are worked out separately, it is assumed that the systematic uncertainties due to the FSR corrections on the W^+ and W^- cross-sections, $\Delta\sigma^{FSR}_{W^+ \to \mu^+ \nu_\mu}$ and $\Delta\sigma^{FSR}_{W^- \to \mu^- \bar{\nu}_\mu}$, are uncorrelated. As a result, the systematic uncertainty on R_{+-} due to the FSR correction, ΔR^{FSR}_{+-}, is

$$\Delta R^{FSR}_{+-} = \sqrt{\left(\frac{\Delta\sigma^{FSR}_{W^+ \to \mu^+ \nu_\mu}}{\sigma_{W^+ \to \mu^+ \nu_\mu}}\right)^2 + \left(\frac{\Delta\sigma^{FSR}_{W^- \to \mu^- \bar{\nu}_\mu}}{\sigma_{W^- \to \mu^- \bar{\nu}_\mu}}\right)^2} \cdot R_{+-} \tag{5.25}$$

As the muon track detector efficiencies are taken as the charge unbiased efficiencies in the W cross-section calculation, and this charge unbiased efficiency is a weighted average of the detector efficiencies of positive muon and negative muon, the systematic uncertainties on the W^+ and W^- cross-sections due to the detector efficiency determination are correlated. With the same reason, the systematic uncertainties on the W^+ and W^- cross-sections due to the selection or acceptance efficiency determination are correlated.

Since the ratio between the positive muon and negative muon in the decay in flight sample is constrained to the ratio in the TIS tracks, the systematic uncertainties on the W^+ and W^- cross-sections due to the background estimation are also correlated. The same argument is valid for the systematic uncertainty on the W cross-section due to the background normalizations, as the ratios of positive and negative muons in the $Z \to \mu\mu$ and $W \to \tau\nu_\tau$ samples are constrained.

As a summary, the systematic uncertainties on the W^+ and W^- cross-sections due to sources which do not include the FSR correction[❶], $\Delta\sigma^{syst\text{-not-}FSR}_{W^+ \to \mu^+ \nu_\mu}$ and $\Delta\sigma^{syst\text{-not-}FSR}_{W^- \to \mu^- \bar{\nu}_\mu}$, are assumed to be correlated. The systematic uncertainty on R_{+-} due to sources which do not include the FSR correction, $\Delta R^{syst\text{-not-}FSR}_{+-}$, then is determined as

$$\Delta R^{syst\text{-not-}FSR}_{+-} = \left| \frac{\Delta\sigma^{syst\text{-not-}FSR}_{W^+ \to \mu^+ \nu_\mu}}{\sigma_{W^+ \to \mu^+ \nu_\mu}} - \frac{\Delta\sigma^{syst\text{-not-}FSR}_{W^- \to \mu^- \bar{\nu}_\mu}}{\sigma_{W^- \to \mu^- \bar{\nu}_\mu}} \right| \cdot R_{+-} \tag{5.27}$$

---

❶ This systematic uncertainty on the W cross-section is determined by adding the uncertainties due to sources which do not include the FSR correction in quadrature

The systematic uncertainty on R_{+-} due to all sources described in section 5.11.3, $\Delta R_{+-}^{syst.}$, is determined by adding ΔR_{+-}^{FSR} and $\Delta R_{+-}^{syst\text{-}not\text{-}FSR}$ in quadrature, and it is written as follows:

$$\Delta R_{+-}^{syst.} = \sqrt{(\Delta R_{+-}^{FSR})^2 + (\Delta R_{+-}^{syst\text{-}not\text{-}FSR})^2} \tag{5.28}$$

The uncertainties on the W^+ and W^- cross-sections due to the luminosity determination, $\Delta\sigma_{W^+\to\mu^+\nu_\mu}^{lumi.}$ and $\Delta\sigma_{W^-\to\mu^-\bar{\nu}_\mu}^{lumi.}$, are also assumed to be correlated, thus the uncertainty on R_{+-} due to the luminosity determination, $\Delta R_{+-}^{lumi.}$, is

$$\Delta R_{+-}^{lumi.} = \left|\frac{\Delta\sigma_{W^+\to\mu^+\nu_\mu}^{lumi.}}{\sigma_{W^+\to\mu^+\nu_\mu}} - \frac{\Delta\sigma_{W^-\to\mu^-\bar{\nu}_\mu}^{lumi.}}{\sigma_{W^-\to\mu^-\bar{\nu}_\mu}}\right| \cdot R_{+-} \tag{5.29}$$

Table 5.27 shows the ratios between the $W^+ \to \mu^+\nu_\mu$ and $W^- \to \mu^-\bar{\nu}_\mu$ cross-sections in each η bin as well as the whole η range. As described in section 5.11.4, the percentage uncertainty on both W^+ and W^- cross-sections due to the luminosity determination is 3.5%, thus this uncertainty on the cross-section ratio is completely cancelled. However, the systematic uncertainty on the cross-section ratio is not cancelled, as the percentage systematic uncertainties on the W^+ and W^- cross-sections are different (see Tables 5.25 and 5.26). The uncertainties quoted in Table 5.27 are the statistical and systematic uncertainties.

Table 5.27 Ratios between the $W^+ \to \mu^+\nu_\mu$ and $W^- \to \mu^-\bar{\nu}_\mu$ cross-sections in each η bin as well as the whole η range

η	R_{+-}	η	R_{+-}
2.0 — 2.5	1.748 ± 0.052 ± 0.007	3.5 — 4.0	0.594 ± 0.028 ± 0.003
2.5 — 3.0	1.438 ± 0.038 ± 0.001	4.0 — 4.5	0.344 ± 0.056 ± 0.023
3.0 — 3.5	1.038 ± 0.033 ± 0.001	2.0 — 4.5	1.292 ± 0.020 ± 0.002

5.13 $\sigma_{W\to\mu\nu_\mu}$ charge asymmetry

The charge asymmetry of the W boson cross-sections, A_{+-}, is defined as

$$A_{+-} = \frac{\sigma_{W^+\to\mu^+\nu_\mu} - \sigma_{W^-\to\mu^-\bar{\nu}_\mu}}{\sigma_{W^+\to\mu^+\nu_\mu} + \sigma_{W^-\to\mu^-\bar{\nu}_\mu}} \tag{5.30}$$

As A_{+-} can be rewritten in terms of R_{+-} as follows:

$$A_{+-} = 1 - \frac{2}{R_{+-} + 1} \tag{5.31}$$

where R_{+-} is the cross-section ratio. The uncertainty on A_{+-} is

$$\Delta A_{+-} = \frac{2\Delta R_{+-}}{(R_{+-} + 1)^2} \tag{5.32}$$

where ΔR_{+-} is the uncertainty on R_{+-}. Table 5.28 shows the charge asymmetries between the $W^+ \to \mu^+ \nu_\mu$ and $W^- \to \mu^- \bar{\nu}_\mu$ cross-sections in each η bin. The uncertainties quoted are the statistical and systematic uncertainties. The total W cross-section charge asymmetry varies rapidly over the whole η range 2.0-4.5, and it is a more indirect measurement compared to the cross-section ratio, thus it is not shown in the table.

Table 5.28 Charge asymmetries between the $W^+ \to \mu^+ \nu_\mu$ and $W^- \to \mu^- \bar{\nu}_\mu$ cross-sections in each η bin

η	A_{+-}	η	A_{+-}
2.0 — 2.5	0.2852 ± 0.0134 ± 0.0017	3.5 — 4.0	-0.2541 ± 0.0220 ± 0.0026
2.5 — 3.0	0.1778 ± 0.0128 ± 0.0005	4.0 — 4.5	-0.4762 ± 0.0620 ± 0.0249
3.0 — 3.5	0.0191 ± 0.0160 ± 0.0005		

5.14 $\sigma_{W \to \mu \nu_\mu}$ theoretical predictions

In order to test the standard model, the experimental results in sections 5.11, 5.12 and 5.13 are compared with theoretical predictions. These theoretical predictions for the $W^\pm$ cross-sections are calculated in the same fiducial phase-space as those applied in data by DYNNLO at NNLO with the MSTW08, ABKM09 and JR09 PDF sets.

There are two uncertainties on the cross-section predictions: the uncertainty due to PDF errors and the uncertainty due to scale variations. As described in section 2.2.1, the hadronic cross-section is a product of the PDF and partonic cross-section, thus uncertainties on PDFs are propagated to the hadronic cross-section. For a particular PDF set, there are several PDF eigenvectors[110]. For each PDF eigenvector, there is an PDF eigenvalue associated with it. If the positive and negative shifts of the PDF eigenvalue move the cross-section above and below its central value, then the difference between the above (below) value and central value is taken as the upper (lower) uncertainty on the cross-section due to this PDF eigenvalue error. If the positive and negative shifts of the PDF eigenvalue move the cross-section in the same direction, then the maximum of the differences between the cross-sections after shifts and the central value as the symmetric uncertainty of the cross-section due to this PDF eigenvalue error. It is assumed that errors on the eigenvalues are uncorrelated with each other, therefore the total uncertainty on the cross-section prediction due to PDF errors is determined by adding uncertainties due to each PDF eigenvalue error in quadrature (see Tables 5.29, 5.30 and 5.31). The uncertainty on the MSTW08 (ABKM09, JR09) PDF estimated at 68% (90%) confidence level corresponds to 1σ (1.645σ) error. In order to get uncertainties at 68% confidence level, the uncertainties on the ABKM09 and JR09 PDF sets are divided by 1.645. The uncertainty on the cross-section predictions due to scale variations is estimated by varying the renormalization and factorization scales together and separately by a factor 2 and 0.5 around the nominal scale, which is set to the W boson mass. These 6 scale variations give six deviations. The maximum value among these six deviations is taken as the uncertainty due to scale variations (see Table 5.32). As uncertainties due to PDF errors are uncorrelated with uncertainties due to scale variations, the total uncertainties on theoretical cross-section predictions are determined by adding these two uncertainties in quadrature.

Table 5.29 Theoretical predictions for the $W^\pm$ cross-sections, ratios and charge asymmetries with the MSTW08 PDF set in each η bin as well as the whole η range. The uncertainties quoted are due to PDF errors

η	$\sigma_{W^+ \to \mu^+ \nu_\mu}$ (pb)	$\sigma_{W^- \to \mu^- \bar\nu_\mu}$ (pb)	R_{+-}	A_{+-}
2.0 — 2.5	$374.0^{+10.1}_{-9.7}$	$220.0^{+6.4}_{-6.4}$	$1.697^{+0.089}_{-0.054}$	$0.259^{+0.020}_{-0.014}$
2.5 — 3.0	$284.0^{+14.1}_{-14.8}$	$188.0^{+5.9}_{-5.4}$	$1.509^{+0.063}_{-0.116}$	$0.203^{+0.020}_{-0.030}$
3.0 — 3.5	$158.0^{+11.8}_{-12.2}$	$145.0^{+4.3}_{-6.0}$	$1.081^{+0.052}_{-0.097}$	$0.039^{+0.024}_{-0.040}$
3.5 — 4.0	$58.1^{+0.7}_{-0.8}$	$91.5^{+2.4}_{-4.1}$	$0.635^{+0.024}_{-0.021}$	$0.223^{+0.018}_{-0.015}$
4.0 — 4.5	$12.6^{+0.3}_{-0.3}$	$40.1^{+2.4}_{-2.3}$	$0.315^{+0.025}_{-0.015}$	$0.521^{+0.030}_{-0.018}$
2.0 — 4.5	$885.5^{+15.4}_{-15.5}$	$686.3^{+13.7}_{-17.7}$	$1.290^{+0.020}_{-0.043}$	

Table 5.30 Theoretical predictions for the $W^\pm$ cross-sections, ratios and charge asymmetries with the ABKM09 PDF set in each η bin as well as the whole η range. The uncertainties quoted are due to PDF errors

η	$\sigma_{W^+ \to \mu^+ \nu_\mu}$ (pb)	$\sigma_{W^- \to \mu^- \bar\nu_\mu}$ (pb)	R_{+-}	A_{+-}
2.0 — 2.5	387.0 ± 18.6	216.0 ± 6.4	1.798 ± 0.078	0.285 ± 0.020
2.5 — 3.0	290.0 ± 7.0	184.0 ± 3.3	1.576 ± 0.039	0.224 ± 0.012
3.0 — 3.5	162.0 ± 4.3	142.0 ± 3.6	1.150 ± 0.029	0.070 ± 0.013
3.5 — 4.0	62.1 ± 1.9	90.4 ± 4.6	0.689 ± 0.037	-0.184 ± 0.026
4.0 — 4.5	14.6 ± 2.0	41.5 ± 1.4	0.352 ± 0.049	-0.479 ± 0.052
2.0 — 4.5	916.0 ± 15.9	673.9 ± 10.7	1.360 ± 0.017	

Table 5.31 Theoretical predictions for the $W^{\pm}$ cross-sections, ratios and charge asymmetries with the JR09 PDF set in each η bin as well as the whole η range. The uncertainties quoted are due to PDF errors

η	$\sigma_{W^+ \to \mu^+ \nu_\mu}$ (pb)	$\sigma_{W^- \to \mu^- \bar\nu_\mu}$ (pb)	R_+	A_+
2.0 — 2.5	$355.0^{+8.9}_{-11.3}$	$203.0^{+5.1}_{-5.7}$	$1.746^{+0.032}_{-0.028}$	$0.272^{+0.008}_{-0.007}$
2.5 — 3.0	$260.0^{+8.7}_{-7.4}$	$180.0^{+4.6}_{-6.8}$	$1.489^{+0.064}_{-0.035}$	$0.196^{+0.020}_{-0.011}$
3.0 — 3.5	$148.0^{+6.4}_{-4.5}$	$142.0^{+7.6}_{-4.1}$	$1.047^{+0.044}_{-0.044}$	$0.023^{+0.030}_{-0.020}$
3.5 — 4.0	$55.1^{+1.5}_{-2.6}$	$91.4^{+4.2}_{-5.7}$	$0.610^{+0.042}_{-0.035}$	$-0.242^{+0.030}_{-0.015}$
4.0 — 4.5	$12.6^{+0.4}_{-0.8}$	$40.4^{+5.7}_{-2.0}$	$0.313^{+0.019}_{-0.049}$	$-0.523^{+0.022}_{-0.060}$
2.0 — 4.5	$839.6^{+24.0}_{-22.5}$	$657.4^{+20.3}_{-20.9}$	$1.280^{+0.040}_{-0.040}$	

Table 5.32 Uncertainties on the $W^{\pm}$ cross-section, ratio and charge asymmetry predictions due to scale variations in each η bin as well as the whole η range

η	$\Delta\sigma_{W^+ \to \mu^+ \nu_\mu}$ (pb)	$\Delta\sigma_{W^- \to \mu^- \bar\nu_\mu}$ (pb)	R_+	A_+
2.0 — 2.5	$^{+5.9}_{-5.8}$	$0^{+5.3}_{-2.6}$	$^{+0.001}_{-0.040}$	$^{+0.005}_{-0.005}$
2.5 — 3.0	$^{+4.0}_{-5.3}$	$^{+2.0}_{-3.4}$	$^{+0.012}_{-0.009}$	$^{+0.004}_{-0.003}$
3.0 — 3.5	$^{+2.7}_{-3.0}$	$^{+2.5}_{-3.5}$	$^{+0.013}_{-0.007}$	$^{+0.008}_{-0.003}$
3.5 — 4.0	$^{+0.9}_{-1.3}$	$^{+2.0}_{-2.8}$	$^{+0.018}_{-0.001}$	$^{+0.006}_{-0.006}$
4.0 — 4.5	$^{+0.6}_{-0.4}$	$^{+1.6}_{-0.7}$	$^{+0.001}_{-0.008}$	$^{+0.005}_{-0.006}$
2.0 — 4.5	$^{+14.0}_{-15.0}$	$^{+11.3}_{-12.0}$	$^{+0.004}_{-0.001}$	

It is assumed that uncertainties on the W^+ and W^- cross-section predictions due to a particular PDF eigenvector error (a particular scale variation) are correlated, thus the relative uncertainty on the W cross-section ratio due to that eigenvector error (that scale variation) is determined as the difference between the relative uncertainties on W^+ and W^- cross-sections. The total uncertainty on the cross-section ratio due to PDF errors is determined by adding the uncertainties due to each PDF eigenvector error in quadrature. The uncertainty on the cross-section ratio due to scale variations is determined as the maximum values among the uncertainties due to each scale variation. The uncertainties on the cross-section charge asymmetry prediction due to PDF errors and scale variations are estimated in a similar way as the uncertainties on the ratio.

Figure. 5.36 shows the differential $W^{\pm}$ cross-sections. These differential cross-sections have been divided by the bin size, 0.5. The uncertainty on the data result is calculated by adding the statistical uncertainty, systematic uncertainty and uncertainty due to the luminosity determination in quadrature. The data results are well described by the MSTW08 theoretical predictions. The ABKM09 theoretical predictions slightly overshoot the W^+ cross-section in the third η bin and the W^- cross-section in the fifth η bin, and slightly undershoot the W^- cross-section in the second η bin. The JR09 theoretical predictions for $W^\pm$ cross-sections are slightly underestimated in the first and second η bins.

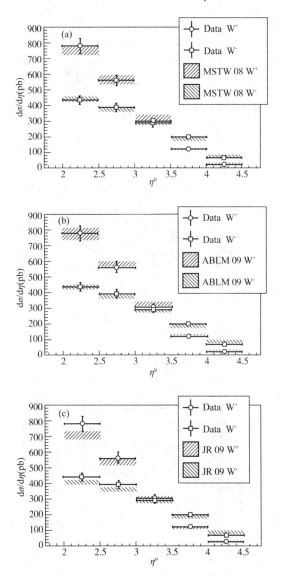

Figure 5.36 Differential $W^\pm$ cross-sections with the data results and theoretical predictions. Data results are presented as points with error bars. (a), hatched areas are MSTW08 predictions. (b), hatched areas are ABKM09 predictions. (c), hatched areas are JR09 predictions

Figure. 5.37 (5.38) shows the differential cross-section ratios (charge asymmetries) of data results and theoretical predictions. The MSTW08 and JR09 theoretical predictions well describe the cross-section ratios (charge asymmetries) in data. The ABKM09 theoretical predictions are overestimated in the second, third and fourth η bins.

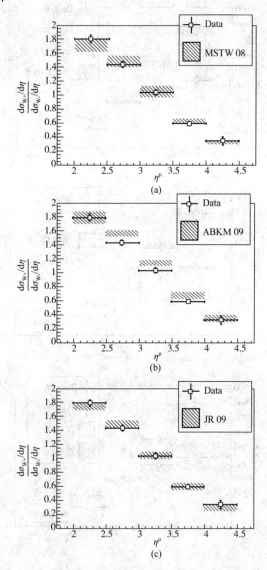

Figure 5.37 Differential cross-section ratios with the data results and theoretical predictions. Data results are presented as points with error bars. (a), hatched areas are MSTW08 predictions. (b), hatched areas are ABKM09 predictions. (c), hatched areas are JR09 predictions

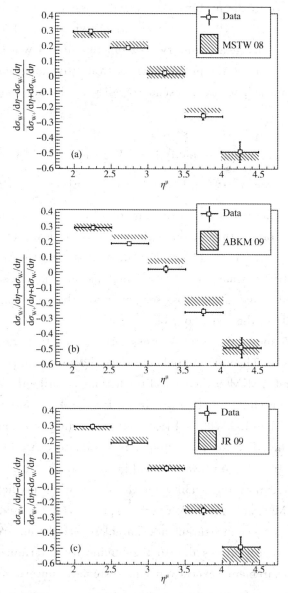

Figure 5.38 Differential cross-section charge asymmetries with the data results and theoretical predictions. Data results are presented as points with error bars. (a), hatched areas are MSTW08 predictions. (b), hatched areas are ABKM09 predictions. (c), hatched areas are JR09 predictions

As described in section 2.2.4 (2.2.5), the cross-section ratio (charge asymmetry) provides a good test of valence u and d quarks. The cross-section ratio shows the ratio between these two quarks while the charge asymmetry shows the difference between these two quarks. As there are two valence u quarks and only one valence d quark in the proton, the overall cross-section ratio (charge asymmetry) is greater than 1 (0). However, as shown in Eq. 2.103 (2.97), the cross-section ratio (charge asymmetry) is a function of θ^*, which is the polar angle between the muon and the proton beam with positive longitudinal momentum in the W rest frame. The muon pseudo-rapidity in the lab frame, η^μ, can be written as[①]

$$\eta^\mu \approx \eta^W + \eta^* \tag{5.33}$$

where η^* is the muon pseudo-rapidity in the W rest frame, η^W is the W pseudo-rapidity in the lab frame, $\eta^W > 0$ in LHCb. As $\eta^* = -\ln\left[\tan\dfrac{\theta^*}{2}\right]$, the cross-section ratio (charge asymmetry) is also a function of η^*. At high muon pseudo-rapidities in the W rest frame (thus at the high pseudo-rapidities in the lab frame), θ^* is small and $(1+\cos\theta^*)^2 \gg (1-\cos\theta^*)^2$. As a result, the cross-section ratio (charge asymmetry) is small than 1 (0) in high pseudo-rapidities. This is exactly what we observed in data: the cross-section ratio (charge asymmetry) becomes smaller than 1 (0) in the last two η bins.

Figure. 5.39 shows the total W cross-sections and their ratio. Data results are shown as bands. Theoretical predictions are shown as points with error bars. The MSTW08 and ABKM09 theoretical predictions describe the $W^\pm$ cross-section measurements well while the JR09 prediction for the W^+ cross-section is slightly underestimated. The MSTW08 and JR09 ratio predictions are consistent with data results while the ABKM09 prediction is overestimated. As the precision of the total W^+ (W^-) cross-section measurement is about 4.7% (4.6%), which is larger than the percentage uncertainty on the W^+ (W^-) prediction due to the MSTW08, ABKM09 and JR09 PDF errors: 1.8%, 1.7% and 2.9% (2.6%, 1.6% and 3.2%), these measurements can not reduce the uncertainty on PDFs. However, as the uncertainties due to the luminosity determination on the $W^\pm$ cross-section measurements cancel out and systematic uncertainties are correlated, the dominant uncertainty on the cross-section ratio measurement is statistical. The precision of the total cross-section ratio measurement is 1.6%, which is smaller than the percentage uncertainty on the ratio prediction due to the MSTW08 and JR09 PDF errors: 3.3% and 3.1%. As a result, measurements on

---

① When the particle travels close to the speed of light, or the mass of the particle is close to zero, the particle's pseudo-rapidity is close to its rapidity. Thus the equation for pseudo-rapidity in Eq. (5.33) is similar as the equation for rapidity in Eq. (2.92).

the W cross-section ratio can reduce the uncertainties on the predictions due to the MSTW08 and JR09 PDF errors.

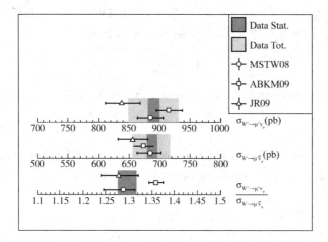

Figure 5.39 Total W cross-sections and their ratio. Data results are shown as bands. Theoretical predictions are shown as points with error bars

5.15 Comparing LHCb results with ATLAS

In this section, we will only compare the W cross-section and its charge asymmetry measurements at LHCb in this book with the results at ATLAS. As explained in section 2.5, in Ref. [47] the correction factors for extrapolations of the LHCb results from $p_T^l = 20$ GeV/c to 25 GeV/c and 30 GeV/c are not available, thus we do not compare the W cross-section and its charge asymmetry measurements in LHCb with the results in CMS.

The datasets and fiducial phase-spaces for the W cross-section and its charge asymmetry measurements in LHCb and ATLAS are described in section 2.5. As a summary, we tabulate the fiducial phase-spaces utilized by LHCb in this book and ATLAS in Table 5.33.

Table 5.33 Fiducial phase-spaces for the measurements performed by LHCb in this book and ATLAS

Measurement	p_T^l (GeV/c)	η^l	M_T (GeV/c^2)	p_T^ν (GeV/c)
LHCb	>20	$2.0 < \eta^l < 4.5$		
ATLAS	>20	$|\eta^l| < 2.5$	>40	>25

As described in section 2.5, for the W cross-section the correction factor in a given i^{th} η bin, C_i, is defined as the predicted cross-section in the fiducial phase-space of ATLAS, σ_i^{ATLAS}, divided by the predicted W cross-section for the fiducial phase-space in LHCb, σ_i^{LHCb}, and is written as follows[47]:

$$C_i = \frac{\sigma_i^{\text{ATLAS}}}{\sigma_i^{\text{LHCb}}} \quad (5.34)$$

Then the extrapolated W cross-section in LHCb, $\sigma_i^{\text{extrapol}}$, is determined as

$$\sigma_i^{\text{extrapol}} = \sigma_i^{\text{measured}} \cdot C_i = \frac{\sigma_i^{\text{ATLAS}}}{\sigma_i^{\text{LHCb}}} \quad (5.35)$$

where $\sigma_i^{\text{measured}}$ is the measured W cross-section in the LHCb fiducial phase-space. For the asymmetry, the correction factor in a given i^{th} η bin, D_i, is defined as[47]

$$D_i = A_i^{\text{ATLAS}} - A_i^{\text{LHCb}} \quad (5.36)$$

where A_i^{ATLAS} (A_i^{LHCb}) is the predicted W cross-section asymmetry in the fiducial phase-space of ATLAS (LHCb). The extrapolated W cross-section charge asymmetry, A_i^{extrapol}, then is determined as

$$A_i^{\text{extrapol}} = A_i^{\text{measured}} + D_i = A_i^{\text{measured}} + A_i^{\text{ATLAS}} - A_i^{\text{LHCb}} \quad (5.37)$$

where A_i^{measured} is the measured W cross-section charge asymmetry in the LHCb fiducial phase-space. In Eqs. (5.34) and (5.36), the predicted W cross-sections and their charge asymmetries in LHCb and ATLAS are estimated by FEWZ❶[111] at NLO with the MSTW08 PDF set. There are several uncertainties associated to the correction factor, and these uncertainties are described in detail in Ref.[47]. The total uncertainty on the correction factor is determined by adding these uncertainties in quadrature. Here we only tabulate the correction factors and their total uncertainties for the W^+, W^- cross-sections and their charge asymmetry in Tables 5.34, 5.35 and 5.36.

Table 5.34 Correction factors for extrapolating the W^+ cross-sections from the LHCb fiducial phase-space ($p_T^l > 20$ GeV/c) to the ATLAS fiducial phase-space ($p_T^l > 20$ GeV/c, $p_T^\nu > 20$ GeV/c and $M_T > 40$ GeV/c^2). Taken from Ref.[47]

η	C_i	Total error	η	C_i	Total error
2.0 — 2.5	0.844	+0.004 / −0.007	3.5 — 4.0	0.868	+0.009 / −0.008
2.5 — 3.0	0.884	+0.004 / −0.003	4.0 — 4.5	0.790	+0.013 / −0.014
3.0 — 3.5	0.901	+0.006 / −0.005			

---

❶ Short for "Fully Exclusive W and Z production"

Table 5.35 Correction factors for extrapolating the W^- cross-sections from the LHCb fiducial phase-space ($p_T^l > 20$ GeV/c) to the ATLAS fiducial phase-space ($p_T^l > 20$ GeV/c, $p_T^\nu > 20$ GeV/c and $M_T > 40$ GeV/c^2). Taken from Ref.[47]

η	C_i	Total error	η	C_i	Total error
2.0 — 2.5	0.831	+0.004 / −0.005	3.5 — 4.0	0.732	+0.011 / −0.012
2.5 — 3.0	0.818	+0.006 / −0.007	4.0 — 4.5	0.625	+0.018 / −0.015
3.0 — 3.5	0.791	+0.008 / −0.008			

Table 5.36 Correction factors for extrapolating the W cross-section charge asymmetry from the LHCb fiducial phase-space ($p_T^l > 20$ GeV/c) to the ATLAS fiducial phase-space ($p_T^l > 20$ GeV/c, $p_T^\nu > 20$ GeV/c and $M_T > 40$ GeV/c^2). Taken from Ref.[47]

η	D_i	Total error	η	D_i	Total error
2.0 — 2.5	0.0071	+0.0014 / −0.0015	3.5 — 4.0	0.0823	+0.0061 / −0.0061
2.5 — 3.0	0.0367	+0.0028 / −0.0028	4.0 — 4.5	0.0916	+0.0086 / −0.0084
3.0 — 3.5	0.0652	+0.0035 / −0.0036			

Figure. 5.40 shows the comparison between the extrapolated $W^\pm$ cross-section in LHCb and the measured $W^\pm$ cross-section in ATLAS. Figure. 5.41 shows the comparison between the extrapolated W cross-section charge asymmetry in LHCb and the measured W cross-section charge asymmetry in ATLAS. The W cross-section and its charge asymmetry results in ATLAS are taken from Ref. [44] and they are estimated with combing the muon and electron channels together. In the overlap region $2.0 < \eta < 2.5$, the extrapolated W^+ cross-sections in LHCb slightly overshoot the measured results in ATLAS, while the extrapolated W^- cross-sections and the W cross-section charge asymmetries agree with the measured results in ATLAS well. In the forward region $2.5 < \eta < 4.5$, the W cross-section and its charge asymmetry measurements in LHCb serve as a nice complement to the ATLAS results.

As a crosscheck, the measured W cross-section and its charge asymmetry results in this book are also compared to the LHCb publication results in Ref. [51]. As the fiducial phase-space applied in this book is the same as that applied in the LHCb publication, we can directly compare the results in the book with the results in the publication without applying the correction factors for extrapolation. Figure. 5.42 shows the com-

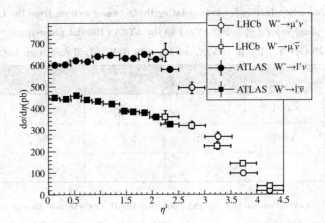

Figure 5.40 Comparison between the extrapolated $W^\pm$ cross-section in LHCb and the measured $W^\pm$ cross-section in ATLAS. The black (white) dots with error bars show the measured (extrapolated) $W^\pm$ cross-sections in ATLAS (LHCb)

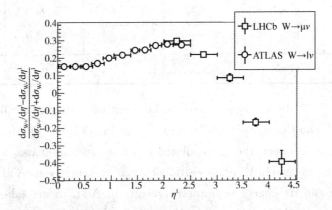

Figure 5.41 Comparison between the extrapolated W cross-section charge asymmetry in LHCb and the measured W cross-section charge asymmetry in ATLAS. The circle (square) dots with error bars show the measured (extrapolated) W cross-section charge asymmetries in ATLAS (LHCb)

parison between the W cross-sections in the book and publication. Figure. 5.43 shows the comparison between the W cross-section charge asymmetries in the book and publication. The W cross-section results in the book slightly overshoot the results in the publication in the first and second η bins while they agree well with the publication results in the third, fourth and fifth η bins. The discrepancy between the results in the book and publication could be mainly due to the following reason: in the publication, there is a bug in the code that determines the track reconstruction efficiency (see section 5.7.

2). This bug makes the measured track reconstruction efficiencies in the publication about 4% higher than their correct efficiencies at the first and second η bins. As a result, after fixing the bug, the cross-sections at the first and second η bins in the publication should be shifted towards higher values. Since the bug changes the positive and negative track reconstruction efficiencies in the first and second η bins at the same percentage level, the W cross-section charge asymmetries at the first and second η bins in the publication looks consistent with the results in the book.

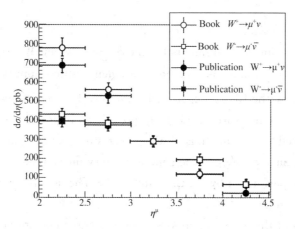

Figure 5.42 Comparison between the W cross-sections in the book and publication. The white (black) dots with error bars show the results in the book (publication)

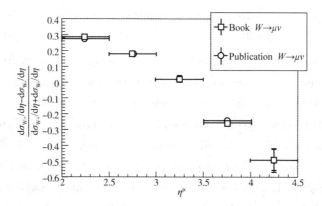

Figure 5.43 Comparison between the W cross-section charge asymmetries in the book and publication. The square (circle) dots with error bars show the results in the book (publication)

Chapter 6

Conclusion

The cross-section measurements for the $W^+ \to \mu^+ \nu_\mu$ and $W^- \to \mu^- \bar{\nu}_\mu$ processes have been detailed in this book. The measurements utilize approximately 37 pb^{-1} of data collected by the LHCb experiment at $\sqrt{s} = 7$ TeV in the year of 2010. The motivations of these measurements are to provide an important test of the standard model as well as a reduction of the uncertainty on the parton distribution function. Additionally they provide complementary measurements of electroweak physics to those performed by ATLAS and CMS. The details for the motivations can be found in section 2.2.6.

The details of the $W \to \mu \nu_\mu$ cross-section measurement can be found in chapter 5. A $W \to \mu \nu_\mu$ event selection scheme is determined with the application of the $W \to \mu \nu_\mu$ simulation sample. The details for the candidate selection scheme can be found in section 5.4. This scheme is applied to the data sample and it yields 26891 candidates for the $W \to \mu \nu_\mu$ process. Of these candidates, 15030 are for the $W^+ \to \mu^+ \nu_\mu$ process and 11861 are for the $W^- \to \mu^- \bar{\nu}_\mu$ process. A fit is performed to determine the purity of muons from the W bosons. This purity is found to be about 79%. A $Z \to \mu\mu$ data sample is utilized to determine efficiencies to reconstruct and select signal events. The details to determine these efficiencies can be found in sections 5.7, 5.8 and 5.9. Finally the cross-section for the $W \to \mu \nu_\mu$ process is calculated in a fiducial phase-space in which the muon transverse momentum is greater than 20 GeV/c and the pseudo-rapidity of the muon is in the range of $2.0 < \eta < 4.5$. The results of the total cross-sections in the fiducial phase-space for the $W^+ \to \mu^+ \nu_\mu$ and $W^- \to \mu^- \bar{\nu}_\mu$ processes are

$$\sigma_{W^+ \to \mu^+ \nu_\mu} = 890.1 \pm 9.2 \pm 26.4 \pm 30.9 \text{pb} \tag{6.1}$$

$$\sigma_{W^- \to \mu^- \bar{\nu}_\mu} = 687.0 \pm 8.1 \pm 19.6 \pm 23.7 \text{pb} \tag{6.2}$$

where the first uncertainty is statistical, the second uncertainty is systematic and the third uncertainty is due to the luminosity determination. The ratio between the total cross-sections for the $W^+ \to \mu^+ \nu_\mu$ and $W^- \to \mu^- \bar{\nu}_\mu$ processes is

$$\frac{\sigma_{W^+ \to \mu^+ \nu_\mu}}{\sigma_{W^- \to \mu^- \bar{\nu}_\mu}} = 1.292 \pm 0.020 \pm 0.002 \tag{6.3}$$

where the first uncertainty is statistical, the second uncertainty is systematic. The uncertainty due to the luminosity determination is completely cancelled.

The measurements of the W cross-sections and their ratio are consistent with theoretical predictions at NNLO with the MSTW08 and JR09 PDF sets. The cross-section ratio prediction with the ABKM09 PDF set is overestimated. The precision of the total W^+ (W^-) cross-section measurement is about 4.7% (4.6%), which is larger than the percentage uncertainty on the W^+ (W^-) prediction due to the MSTW08, ABKM09 and JR09 PDF errors: 1.8%, 1.7% and 2.9% (2.6%, 1.6% and 3.2%). Thus, with current statistics, these measurements can not reduce the uncertainty on PDFs. However, as the uncertainties due to the luminosity determination on the $W^\pm$ cross-section measurements cancel out and systematic uncertainties are correlated, the dominant uncertainty on the cross-section ratio measurement is statistical. The precision of the total cross-section ratio measurement is 1.6%, which is smaller than the percentage uncertainty on the ratio prediction due to the MSTW08 and JR09 PDF errors: 3.3% and 3.1%. As a result, the measurement on the W cross-section ratio can reduce the uncertainties on the predictions due to the MSTW08 and JR09 PDF errors.

As the dominant uncertainty on the total W cross-section is from the integrated luminosity determination, the precision on the W cross-section measurement can be improved by a more precise integrated luminosity measurement. In addition, the systematic and statistical uncertainties on the W cross-section can be reduced by a larger dataset, such as the 2011 (2012) dataset with an integrated luminosity of 1.1 fb^{-1} (2.1 fb^{-1})❶.

---

❶ The results with the 2011 dataset can be found at the thesis CERN-THESIS-2013-178.

Bibliography

[1] Martin B R, Shaw G. Particle Physics[M]. 3rd ed. UK: A John Wiley and Sons, Ltd, Publication, 2008.

[2] Peskin M E, Schroeder D V. An introduction to Quantum Field Theory[M]. USA: Westview Press, 1995.

[3] Feynman R P. Quantum Electrodynamics[M]. USA: Westview Press, 1998.

[4] Schwinger J. On Quantum-Electrodynamics and the Magnetic Mom- ent of the Electron[J]. Physical Review, 1948, 73, 416.

[5] Lamb W E Jr., Retherford R C. Fine Structure of the Hydrogen Atom by a Microwave Method[J]. Physical Review, 1947, 72, 241.

[6] Georgi H, Glashow S L. Unified Weak and Electromagnetic Interac- tions without Neutral Currents[J]. Physical Review Letter, 1972, 28, 1494.

[7] Ellis R K, Stirling W J, Webber B R. QCD and Collider Physics[M] UK: Cambridge University Press, 1996.

[8] Cottingham W N, Greenwood D A. An Introduction to the Standard Model of Particle Physics[M], UK: Cambridge University Press, 2007.

[9] Abe F, et al. Measurement of the ratio $B(W \to \tau\nu)/B(W \to e\nu)$ in $p\bar{p}$ collisions at $\sqrt{s} = 1.8 \text{TeV}$[J]. Physical Review Letter, 1992, 68, 3398.

[10] The UA1 Collaboration. Experimental observation of isolated large transverse energy electrons with associated missing energy at $\sqrt{s} = 540 GeV$[J]. Physics Letter B, 1983, 122, 103-116.

[11] Wilczek F. The cosmic asymmetry between matter and antimatter[J]. Scientific American, 1980, 243, 82-90.

[12] CERN. The Large Electron-Positron Collider[EB/OL]. [2021-03-03]. https://home.cern/science/accelerators/large-electron-positron-collider.

[13] Fermilab. Tevatron[EB/OL]. [2021-03-03]. http://www.fnal.gov/pub/science/experiments/energy/tevatron.

[14] CERN. The Large Hadron Collider[EB/OL]. [2021-03-03]. http://home.web.cern.ch/about/accelerators/large-hadron-collider.

[15] Beringer J, et al. Review of Particle Physics[J]. Physical Review D, 2012, 86, 010001.

[16] The ATLAS Collaboration. Observation of a new particle in the search for the Standard Model Higgs boson with the ATLAS detector at the LHC[J]. Physics Letter B, 2012, 716, 1-29.

[17] The CMS Collaboration. Observation of a new boson at a mass of 125 GeV with the CMS experiment at the LHC[J]. Physics Letter B, 2012, 716, 30-61.

[18] O'Luanaigh C. New results indicate that new particle is a Higgs boson[EB/OL]. [2021-03-03]. http://home.web.cern.ch/about/updates/20-13/03/new-results-indicate-new-particle-higgs-boson.

[19] The NA48 Collaboration. A new measurement of direct CP violation in two pion decays of the neutral kaon[J]. Physics Letter B, 1999, 465, 335-348.

[20] Wikipedia. CP violation [EB/OL]. [2013-09-01]. http://en.wikipedia.org/wiki/CP_violation.

[21] Glashow S L. Partial-symmetries of weak interactions[J]. Nuclear Physics, 1961, 22, 579.

[22] Martin A D, Stirling W J, Thorne R S, Watt G. Parton distributions for the LHC[J], European Physical Journal C, 2009, 63, 189.

[23] Alekhin S, Blumlein J, Klein S, Moch S. 3-, 4-, and 5-flavour next-to-next-to-leading order parton distribution functions from deep-inelastic-scattering data and at hadron colliders[J]. Physical Review D, 2010, 81, 014032.

[24] Jimenez-Delgado P, Reya E. Dynamical NNLO parton distribu- tions[J]. Physical Review D, 2009, 79, 074023.

[25] Aaron F D, et al. Combined measurement and QCD analysis of the inclusive $e^{\pm}p$ scattering cross-sections at HERA[J]. Journal of High Energy Physics, 2010, 01, 109.

[26] Ball R D, et al. A first unbiased global NLO determination of parton distributions and their uncertainties[J]. Nuclear Physics B, 2010, 838, 136.

[27] Nadolsky P M, et al. Implications of CTEQ global analysis for collider observables[J]. Physical Review D, 2008, 78, 013004.

[28] Aitchison I J R, Hey A J G. Gauge Theories in Particle Physics[M]. 2rd edition. UK: Institute of Physics Publishing, 2001.

[29] Stirling W J. Understanding the behavior of the LHC lepton charge asymmetry[R]. Private communication, 2012.

[30] Farry S. PDF Studies at LHCb, ATLAS and CMS, LHCb-TALK- 2013-279[R]. Geneva: CERN, 2013.

[31] Thorne R S, Martin A D, Stirling W J, Watt G. Parton Distributions and QCD at LHCb[EB/OL]. [2021-03-03]. https://arxiv.org/abs/0808.

1847.

[32] Campbell J, Ellis K. MCFM-Monte Carlo for FeMtobarn processes [EB/OL]. [2021-03-03]. http://mcfm.fnal.gov.

[33] Lorenzi F D. Parton Distribution Function Studies and a Measure- ment of Drell-Yan Produced Muon Pairs at LHCb, CERN-THESIS- 2011-237[D]. Geneva: CERN, 2011.

[34] Bethke S. The 2009 world average of α_s[J]. European Physical Journal C, 2009, 64, 689.

[35] Catani S, Ferrera G, Grazzini M. W boson production at hadron colliders: the lepton charge asymmetry in NNLO QCD[J]. Journal of High Energy Physics, 2010, 05, 006.

[36] Sjostrand T, Mrenna S, Skands P. PYTHIA 6.4 Physics and Manual [J]. Journal of High Energy Physics, 2006, 05, 026.

[37] Alioli S, Nason P, Oleari C, Re E. NLO vector-boson production matched with shower in POWHEG[J]. Journal of High Energy Physics, 2008, 07, 060.

[38] Andersson B, Gustafson G, Ingelman G, Sjostrand T. Parton frag- menta-tion and string dynamics[J]. Physics Report, 1983, 97, 31-145.

[39] Gottschalk T D. An improved description of hadronization in the QCD cluster model for e^+e^- annihilation[J]. Nuclear Physics B, 1984, 239, 349-381.

[40] The CDF Collaboration. First Measurements of Inclusive W and Z Cross Sections from Run II of the Fermilab Tevatron Collider[J]. Physi- cal Review Letter, 2005, 94, 091803.

[41] The CDF Collaboration. Direct Measurement of the W Production Charge Asymmetry in $p\bar{p}$ Collisions at $\sqrt{s}=1.96$ TeV[J]. Physical Review Letter, 2019, 102, 181801.

[42] The CDF Collaboration. Measurement of the forward-backward charge a-symmetry from $W \to e\nu$ production in $p\bar{p}$ collisions at $\sqrt{s}=1.96$ TeV[J]. Physical Review D, 2005, 71, 051104(R).

[43] The D0 Collaboration. Measurement of the muon charge asymmetry from W boson decays[J]. Physical Review D, 2008, 77, 011106(R).

[44] The ATLAS Collaboration. Measurement of the inclusive $W^\pm$ and Z/γ^* cross-sections in the e and μ decay channels in pp collisions at $\sqrt{s}=7$ TeV with the ATLAS detector[J]. Physical Review D, 2012, 85, 072004.

[45] The CMS Collaboration. Measurement of the inclusive W and Z production cross-sections in pp collisions at $\sqrt{s}=7$ TeV with the CMS experiment[J]. Journal of High Energy Physics, 2011, 10, 132.

[46] The CMS Collaboration. Measurement of the lepton charge asym- metry in

inclusive W production in pp collisions at $\sqrt{s} = 7$ TeV[J]. Journal of High Energy Physics, 2011, 04, 050.

[47] Muller K. Graphical comparison of the LHCb measurements of W and Z production with ATLAS and CMS, LHCb-ANA-2013-020[R]. Geneva: CERN, 2013.

[48] Terning J. Modern Supersymmetry[M]. UK: Oxford Science Publ-ications, 2006.

[49] The LHCb Collaboration. The LHCb Detector at the LHC[J]. Journal of Instrumentation, 2008, 3, S08005.

[50] The LHCb Collaboration. First Evidence for the Decay $B_s^0 \to \mu^+\mu^-$[J]. Physical Review Letter, 2013, 110, 021801.

[51] The LHCb Collaboration. Inclusive W and Z production in the forward region at $\sqrt{s} = 7$ TeV[J]. Journal of High Energy Physics, 2012, 06, 58.

[52] The LHCb Collaboration. A study of the Z production cross-section in pp-collisions at $\sqrt{s} = 7$ TeV using tau final states[J]. Journal of High Energy Physics, 2013, 01, 111.

[53] The LHCb Collaboration. Measurement of the cross-section for $Z \to e^+e^-$ production in pp collisions at $\sqrt{s} = 7$ TeV[J]. Journal of High Energy Physics, 2013, 02, 106.

[54] Jowett J M. Collision Schedules and Bunch Filling Schemes in the LHC, LHC-PROJECT-NOTE-179[R]. Geneva: CERN, 1999.

[55] The LHC. LHC commissioning with Beam[EB/OL]. [2021-03-03]. http://lhc-commissioning.web.cern.ch/lhc-commissioning/.

[56] Richmond M. The LHC is not going to destroy the Earth[EB/OL]. [2021-03-03]. http://spiff.rit.edu/richmond/asras/lhc/lhc.html.

[57] Herr W, Muratori B. Proceedings of CERN Accelerator School: Concept of luminosity[C], Geneva: CERN, 2003.

[58] Van der Meer S. Calibration of the Effective Beam Height in the ISR, CERNISR-PO-68-31[R]. Geneva: CERN, 1968.

[59] The LHCb Collaboration. Absolute luminosity measurements with the LHCb detector at the LHC[J]. Journal of Instrumentation, 2012, 7, P01010.

[60] The LHCb Collaboration. LHCb-Large Hadron Collider beauty experiment[EB/OL]. [2021-03-03]. http://lhcb-public.web.cern.ch.

[61] The LHCb Collaboration. LHCb magnet: Technical Design Report, CERN-LHCC-2000-007[R]. Geneva: CERN, 2000.

[62] The LHCb Silicon Tracker. LHCb Silicon Tracker-Material for Publications[EB/OL]. [2021-03-03]. http://lhcb.physik.uzh.ch/ST/public/

mat- erial/index. php.
[63] The LHCb Collaboration. LHCb Technical Design Report: Reoptim- ized Detector Design and Performance, CERN-LHCC-2003-030[R]. Geneva: CERN, 2003.
[64] Adinolfi M, et al.. Performance of the LHCb RICH detector at the LHC[J]. European Physical Journal C, 2013, 73, 2431.
[65] Machikhiliyan I. The LHCb Calorimeter System[J]. Nuclear Physics B-Proceedings Supplements, 2008, 177, 178.
[66] Callot O. Improved robustness of the VELO tracking, LHCb-2003- 017[R]. Geneva: CERN, 2003.
[67] Hutchcroft D. VELO Pattern Recognition, LHCb-2007-013[R]. Gen-eva: CERN, 2007.
[68] Callot O, Hansmann-Menzemer S. The forward tracking: Algorithm and performance studies, LHCb-2007-015[R]. Geneva: CERN, 2007.
[69] Van Tilburg J, Merk M. Track simulation and reconstruction in LHCb, CERN-THESES-2005-040[D]. Geneva: CERN, 2005.
[70] Forty R. Track seeding, LHCb-2001-109[R]. Geneva: CERN, 2001.
[71] Needham M, van Tilburg J. Performance of the track matching, LHCb-2007-020[R]. Geneva: CERN, 2007.
[72] Anderson J. Testing the electroweak sector and determining the absolute luminosity at LHCb using dimuon final states, CERN-THESES- 2009-020 [D]. Geneva: CERN, 2009.
[73] Xie Y. Short track reconstruction with VELO and TT, LHCb-2003 -100[R]. Geneva: CERN, 2003.
[74] Callot O. Downstream Pattern Recognition, LHCb-2007-026[R]. Geneva: CERN, 2007.
[75] Krasowski M, et al. Primary vertex reconstruction, LHCb-2007-011[R]. Geneva: CERN, 2007.
[76] Borghi S. Tracking and Alignment Performance of the LHCb silicon detectors, LHCb-PROC-2011-086[R]. Geneva: CERN, 2011.
[77] Forty R, Schneider O. RICH pattern recognition, LHCb-1998-040[R]. Geneva: CERN, 1998.
[78] Farry S. A measurement of Z production using tau final states with the LHCb detector, CERN-THESES-2012-227[D]. Geneva: CERN, 2012.
[79] The LHCb Collaboration. LHCb Calorimeters Technical Design Rep- ort, CERN-LHCC-2000-036[R]. Geneva: CERN, 2000.
[80] Breton V, Brun N, Perret P. A clustering algorithm for the LHCb electromagnetic calorimeter using cellular automaton, LHCb-2001- 123[R]. Ge-

neva: CERN, 2001.

[81] Machefert F. LHCb calorimeter electronics, photon identification, calorimeter calibration[R]. Geneva: CERN, 2011.

[82] Deschamps O, et al. Photon and neutral pion reconstruction, LHCb-2003-091[R]. Geneva: CERN, 2003.

[83] Terrier H, Belyaev I. Particle identification with LHCb calorimeters, LHCb-2003-092[R]. Geneva: CERN, 2003.

[84] The LHCb Calorimeter Group. Performance of the LHCb Calorimet- ers[R]. Geneva: CERN, 2013.

[85] Cid Vidal X. Muon Identification in the LHCb experiment[EB/OL]. [2021-03-02]. https://arxiv.org/abs/1005.2585.

[86] Archilli F, et al. Performance of the Muon Identification at LHCb[J]. Journal of Instrumentation, 2013, 8, P10020.

[87] Lorenzi F D. Parton Distribution Function Studies and a Measurem- ent of Drell-Yan Produced Muon Pairs at LHCb, CERN-THESES-2011-237[D]. Geneva: CERN, 2011.

[88] The LHCb Collaboration. LHCb Trigger System Technical Design Report, CERN-LHCC-2003-031[R]. Geneva: CERN, 2003.

[89] Keaveney J. A measurement of the Z cross section at LHCb, CERN-THESES-2011-202[D]. Geneva: CERN, 2011.

[90] Tuning N. Matching VELO tracks to L0 objects for L1, LHCb-2003-039[R]. Geneva: CERN, 2003.

[91] IN2P3. Cumulative luminosity for 2010 data collection[EB/OL]. [2021-03-03]. http://marwww.in2p3.fr/~legac/LHCb/.

[92] The LHCb Collaboration. LHCb computing: Technical Design Rep- ort, CERN-LHCC-2005-019[R]. Geneva: CERN, 2005.

[93] The LHCb Collaboration. The GAUSS Project[EB/OL]. [2021-03-03]. https://lhcb-comp.web.cern.ch/Simulation/.

[94] The LHCb Collaboration. The BOOLE Project[EB/OL]. [2021-03-03]. https://lhcb-comp.web.cern.ch/Digitization/.

[95] The LHCb Collaboration, The BRUNEL Project[EB/OL]. [2021-03-03]. http://lhcbdoc.web.cern.ch/lhcbdoc/brunel/.

[96] The LHCb Collaboration, The DAVINCI Project[EB/OL]. [2021-03-03]. https://lhcb-comp.web.cern.ch/Analysis/.

[97] Whalley M R, Bourilkov D, Group R C. The Les Houches accord PDFs (LHAPDF) and LHAGLUE[EB/OL]. [2021-03-03]. https://arxiv.org/abs/hep-ph/0508110.

[98] The CTEQ Collaboration. Global QCD analysis of parton structure of the

CTEQ5 parton distributions[J]. European Physical Journal C, 12, 375-392.

Geant 4 Collaboration. Geant 4-a simulation toolkit[J]. Nuclear Instruments and Methods in Physics Research A, 2003, 506, 250303.

Barlow R, Beeston C. Fitting using finite Monte Carlo samples[J]. Computer Physics Communication, 1993, 77, 219-228.

ROOT. TFractionFitter[EB/OL]. [2021-03-03]. https://root.cern/doc/master/classTFractionFitter.html

[102] Vogt R. What is the Real K Factor? [J]. Heavy Ion Physics, 2003, 17, 75.

[103] Watt G. Parton distribution function dependence of benchmark Standard Model total cross sections at the 7 TeV LHC[J]. Journal of High Energy Physics, 2011, 09, 069.

[104] Gaiser J E, et al. Charmonium Spectroscopy from inclusive ψ' and J/ψ radiative decays[J]. Physical Review D, 1986, 34, 711.

[105] Paterno M. Calculating efficiencies and their uncertainties, FERMI-LAB-TM-2286-CD[R]. USA: FERMILAB, 2004.

[106] Eisner A M. Acoplanarity and Beam Line Rotations[EB/OL]. [2021-03-03]. http://www.slac.stanford.edu/BFROOT/www/Detector/Tri-gger/software/L3TFilters/rotation.ps.

[107] Golonka P, Was Z. PHOTOS Monte Carlo: a precision tool for QED corrections in Z and W decays[J]. European Physical Journal C, 2006, 45, 97.

[108] Catani S, Grazzini M. Next-to-Next-to-Leading-Order Subtraction Formalism in Hadron Collisions and its Application to Higgs-Boson Production at the Large Hadron Collider[J]. Physical Review Letter, 2007, 98, 222002.

[109] Catani S, et al. Vector Boson Production at Hadron Colliders: A Fully Exclusive QCD Calculation at Next-to-Next-to-Leading Order[J]. Physical Review Letter, 2009, 103, 082001.

[110] Martin A D, et al. Uncertainties of predictions from parton distributions. I: experimental errors[J]. European Physical Journal C, 2003, 28, 455-473.

[111] Gavin R, et al. FEWZ 2.0: A code for hadronic Z production at next-to-next-to-leading order[J]. Computer Physics Communication, 2011, 182, 2388-2403.

[112] Gao J, et al. The CT10 NNLO Global Analysis of QCD[J]. Physical Review D, 2014, 89, 033009.

[113] The LHCb Collaboration. LHCb Muon System Technical Design Report, CERN-LHCC-2001-010[R]. Geneva: CERN, 2001.